STEAM-H: Science, Technology, Engineering, Agriculture, Mathematics & Health

Series Editor

Bourama Toni, Department of Mathematics, Howard University, Washington, USA

This interdisciplinary series highlights the wealth of recent advances in the pure and applied sciences made by researchers collaborating between fields where mathematics is a core focus. As we continue to make fundamental advances in various scientific disciplines, the most powerful applications will increasingly be revealed by an interdisciplinary approach. This series serves as a catalyst for these researchers to develop novel applications of, and approaches to, the mathematical sciences. As such, we expect this series to become a national and international reference in STEAM-H education and research.

Interdisciplinary by design, the series focuses largely on scientists and mathematicians developing novel methodologies and research techniques that have benefits beyond a single community. This approach seeks to connect researchers from across the globe, united in the common language of the mathematical sciences. Thus, volumes in this series are suitable for both students and researchers in a variety of interdisciplinary fields, such as: mathematics as it applies to engineering; physical chemistry and material sciences; environmental, health, behavioral and life sciences; nanotechnology and robotics; computational and data sciences; signal/image processing and machine learning; finance, economics, operations research, and game theory. The series originated from the weekly yearlong STEAM-H Lecture series at Virginia State University featuring world-class experts in a dynamic forum. Contributions reflected the most recent advances in scientific knowledge and were delivered in a standardized, self-contained and pedagogically-oriented manner to a multidisciplinary audience of faculty and students with the objective of fostering student interest and participation in the STEAM-H disciplines as well as fostering interdisciplinary collaborative research. The series strongly advocates multidisciplinary collaboration with the goal to generate new interdisciplinary holistic approaches, instruments and models, including new knowledge, and to transcend scientific boundaries.

Peer reviewing

All monographs and works selected for contributed volumes within the STEAM-H series undergo peer review. The STEAM-H series follows a single-blind review process. A minimum of two reports are asked for each submitted manuscript. The Volume Editors act in cooperation with the Series Editor for a final decision. The Series Editors agrees with and follows the guidelines published by the Committee on Publication Ethics.

Titles from this series are indexed by Scopus, Mathematical Reviews, and zbMATH.

Edgard A. Pimentel · Bourama Toni
Editors

Differential and Algorithmic Intelligent Game Theory

Methods and Applications

Springer

Editors
Edgard A. Pimentel
Department of Mathematics
CMUC, University of Coimbra
Coimbra, Portugal

Bourama Toni
Department of Mathematics
Howard University
Washington, DC, USA

ISSN 2520-193X ISSN 2520-1948 (electronic)
STEAM-H: Science, Technology, Engineering, Agriculture, Mathematics & Health
ISBN 978-3-031-97732-9 ISBN 978-3-031-97733-6 (eBook)
https://doi.org/10.1007/978-3-031-97733-6

Mathematics Subject Classification: 91A16, 91A22, 91A23, 91A25, 91A68, 91A80

This Springer imprint is published by the registered company Springer Nature Switzerland AG
The registered company address is: Gewerbestrasse 11, 6330 Cham, Switzerland

Preface

Game Theory has developed substantially in recent years through major theoretical breakthroughs as well as using various ingenious applications in socio-economics sciences, agricultural and biomedical sciences, machine/deep learning and artificial intelligence.

This volume, the first instalment of a series on various aspects of Game Theory, features recent developments in Differential and Algorithmic Intelligent Game Theory, its methods and its applications by experts in the field. It serves to re-emphasize the relevance and depth of this important area of mathematics, and, in particular, its expanding reach into the physical, biological, social, and computational sciences, as well as in engineering and technology.

This Volume I and the subsequent volumes cover in depth the dynamic and computationally rich mathematics field at the interface of differential and algorithmic game theory. That is the interface of continuous decision-making processes (differential game) and the algorithmic design principles and mechanisms of game theory, leading to a framework broad and comprehensive to efficiently analyze and solve strategic interactions within complex evolving systems, natural, physical and information-based.

The volumes encompass various segments:

Dynamic strategic interactions where players in the broadest possible sense engage in decision-making over time, respectively continuous, discrete, non-Archimedean and quantized (hypothetically) or quantumized. Strategies are considered under these time scales, including quantum strategies which allow the superposition and entanglement of strategic actions. A combination of techniques including optimal control leads to equilibrium solutions to include solutions á la NASH.

On the other hand, the volumes address the computational challenges of finding equilibria in complex game settings, and feature the development of efficient and competent algorithms to compute Nash equilibria, pure or mixed, subgame or local perfect equilibria.

The intelligent game theory aspects refer to the new and expanding use of game theory to assist artificial intelligence systems decision-making in competitive, cooperative or non-cooperative situations.

These volumes, starting with this first one, highlight the wide and expanding applicability of differential and algorithmically intelligent game theory to various domains, both traditional and non-traditional, such as economics (market competition, resource allocation), engineering (network design, control systems, sensor coverage), computer science (mechanism design, resource allocation in distributed systems), political systems (international relations, policy design), information science (spread of classical and quantum information), quantum computing (quantum game), non-Archimedean infinitesimal games (e.g., randomized strategy games with negative probabilities), evolutionary biology, defense (plane-missile pursuit-evade game, resource allocation warfare, sensing and tracking sensor warfare).

These applications are "modernized" into: AI-assisted Behavioral and Evolutionary game theory. That is, how AI can use game theory principles and algorithms to help understand and predict human and machine behavior in strategic situations. Multi-agent AI systems: indeed, game theoretical methodologies can be used to model interactions between multiple AI systems to derive an equilibrium solution and predict outcomes. Imitation and reinforcement learning. Adversary training in generative adversarial networks. Detecting advanced persistent threats.

The book is published in the Springer series STEAM-H: Science, Technology, Engineering, Agriculture, Mathematics and Health, a pioneer in the field of interdisciplinary publications, fully indexed in the SCOPUS, Google Scholar, MathSciNet, Research Gate database and hosted by Springer at https://link.springer.com/bookseries/15560.

This interdisciplinary series highlights the wealth of recent advances in the pure and applied sciences made by researchers collaborating between fields where mathematics is a core focus. This volume contributes to re-emphasize, if need be, mathematics as the most powerful tool to understand, formulate, address and resolve complex problems.

As we continue to make fundamental advances in various scientific disciplines, the most powerful applications will increasingly be revealed by an interdisciplinary approach. The STEAM-H series catalyzes these researchers to develop novel applications of and approaches to the mathematical sciences. As such, the series is now a national and international reference in STEAM-H education and research.

All manuscripts are peer-reviewed by subject matter experts. The book presents perspectives and development of novel and fundamental ideas for a modern understanding of game-theoretical approaches and modelling. It is indeed an introductory volume for a developing sequence of volumes on Differential and Algorithmic Intelligent Game Theory (D&AIGT). Throughout this volume and the subsequent ones, the practical use of differential and algorithmic game theory is illustrated by real-world case studies and applications, offering a unique perspective by combining the power of differential game theory's continuous decision-making settings with the algorithmic methodology to analyze and design strategic systems in dynamic complex environments.

Indeed, the multidisciplinary and interdisciplinary STEAM-H series brings together leading researchers to present their work in the perspective to advance their specific fields and in a way to generate a genuine interdisciplinary interaction transcending disciplinary boundaries. Contributions are invited only and reflect the most recent advances delivered in a high-standard, self-contained way.

The volume, as well as the subsequent volumes, foster the readership interest and enthusiasm in the STEAM-H disciplines (Science, Technology, Engineering, Agriculture, Mathematics & Health), stimulate graduate and undergraduate research, and generate collaboration between researchers on a genuine interdisciplinary basis, The shared emphasis of these carefully selected and peer-refereed contributed papers is on important methods, research directions, and applications of analysis, modeling and simulations including within and beyond mathematical sciences.

The readership of researchers, graduate and undergraduate students, extends to all science practitioners with a strong foundation in mathematics, computer science, and economics and interested in applying game theory and its subfield of differential and algorithmic game theory to analyze and design strategic systems with dynamic decision-making processes.

The book is featured in the STEAM-H series by now well established as a reference of choice for interdisciplinary scientists and mathematicians and a source of inspiration for a broad spectrum of researchers, with a high impact through all its volumes published by the world-renowned Springer Nature.

Coimbra, Portugal — Edgard A. Pimentel
Washington, DC, USA — Bourama Toni
March 2025

Acknowledgements We would like to express our sincere appreciation to all the contributors and all the anonymous referees for their professionalism. They all made this volume a reality for the greater benefit of the community of science, technology, engineering, agriculture, mathematics and health.

Contents

Contributors

David Evangelista Escola de Matemática Aplicada (EMAp), Fundação Getulio Vargas, Rio de Janeiro, Brazil

Rita Ferreira CEMSE Division, King Abdullah University of Science and Technology (KAUST), Thuwal, Saudi Arabia

Diogo Gomes CEMSE Division, King Abdullah University of Science and Technology (KAUST), Thuwal, Saudi Arabia

Yuri Saporito Escola de Matemática Aplicada (EMAp), Fundação Getulio Vargas, Rio de Janeiro, Brazil

Dantas Serra Departamento de Matemática, Universidade de Coimbra, Coimbra, Portugal

João Luís Soares Departamento de Matemática, Universidade de Coimbra, Coimbra, Portugal

Teruo Tada CEMSE Division, King Abdullah University of Science and Technology (KAUST), Thuwal, Saudi Arabia

Yuri Thamsten Instituto de Matemática e Estatística (IME), Universidade Federal Fluminense, Niterói, Brazil

Bourama Toni Department of Mathematics, Howard University, Washington, DC, USA

Thomas A. Wettergren Naval Undersea Warfare Center, Newport, RI, USA

Compendium of Advances in Game Theory: Classical, Differential, Algorithmic, Non-Archimedean and Quantum Games

Bourama Toni

Abstract This compendium features advances in Game Theory, to include:

- Classical Game Theory: Cooperative and non-cooperative. Zero-sum and non-zero sum games. Potential and Congestion games. Mean Field games. Nash Equilibrium, Correlated Nash Equilibrium and Approximate Nash Equilibrium. Evolutionary Game Theory
- Intelligent Game: Differential Game Theory. Algorithm Game Theory and Security Games.
- Quantum Games and Quantumization of classical games such as the Battle of the Sexes
- Non-Archimedean and p-adic game theory and its growing relevancy as the domains of game-theoretic application expands.
- p-adic quantum game to leverage and combine the distinguishing features of non-Archimedean analysis and quantum information theory. This is a novel game-theoretic approach with great potential of application.

In times of exponential growth of artificial intelligence and machine learning and the dawn of post-human mathematical creativity, this compendium is meant to be a reference of choice for all game theory researchers.

Keywords Game theory · Non-cooperative games · Nash equilibrium · Approximate nash equilibrium · Mean field game · Evolution matrix game · Battle of the sexes · Differential game · Algorithmic game · Security games · Non-archimedean game · Quantum game · Superposition and entanglement · Complex hilbert space · p-adic Hilbert space · p-adic quantum game

---

B. Toni (✉)
Department of Mathematics, Howard University, Washington, DC, USA
e-mail: bourama.toni@howard.edu

E. A. Pimentel and B. Toni (eds.), *Differential and Algorithmic Intelligent Game Theory*, STEAM-H: Science, Technology, Engineering, Agriculture, Mathematics & Health,
https://doi.org/10.1007/978-3-031-97733-6_1

In Memoriam: John F. Nash (1928–2015)

Towards a concept of Nash Limit Cycle, Stony Book July 2013

At the July 2013 International Conference on Game Theory at Stony Brook, we were fortunate to meet and discuss with the late John Nash some of our game-theoretic ideas, in particular, the concept of Nash Limit cycles and its application to socio-cultural evolution. His inputs were valuable to us then, as they are today, and we dedicate this work and what it is worth to his memory and his everlasting contributions to this old age Theory of Games. It has been a privilege to get to know John Nash. His genuine kindness and keen show of interest in our game-theoretical work and the STEAM-H series is fondly remembered: he so kindly autographed our first two early books.

Outline

1. Introduction and Background
2. Evolutionary Game Theory
3. Intelligent Game
4. Differential Game Theory
5. Algorithmic Game Theory and Security Games
6. Quantum Game Theory
7. Non-Archimedean Game Theory
8. p-Adic Quantum Game Theory
9. A socio-cultural game model
10. Concluding Remarks
11. References

1 Introduction and Background

1.1 Raison D'être

Game Theory has evolved from an intuitive to a formal analysis and understanding of interacting entities, be it, organism, animals, programs, social norms, etc. Its underlying ideas and *raison d'être* transpired throughout history: from the biblical game-theoretic story of the Babylonian *Talmud*[1] *Estate Division Problem*[2] to the works of Descartes,[3] Darwin,[4] and the teachings of the Chinese warrior philosopher Sun Tzu [79].[5]

---

[1] The Talmud, more precisely the Babylonian Talmud, comprises of The Mishna (c. 200 CE), a written compendium of Judaism's Oral Law and The Gemara (c. 500 CE), which is a record of discussions by rabbis about the Mishna. It is subdivided into 60 books with the first printed version appearing around 1520. A modern English translation has 73 volumes. The Mishna is in Hebrew, but the Gemara is Aramaic; The Talmud is thought to be largely incomprehensible without the commentary of the French rabbi Rashi.

[2] According to the Babylonian Talmud, a man dies without paying all his three creditors but leaves an estate too small to pay his debts. Creditor 1 is owed 100, Creditor 2 is owed 200 and Creditor 3 is owed 300. If the estate is 100, each creditor gets 33 1/3, If the estate is 200, Creditor 1 gets 50, Creditors 2 and 3 get 75 each, And finally if the estate is 300, Creditor 1 gets 50, Creditor 2 gets 100, and Creditor 3 gets 150. The estate division algorithm is based on the legal system's principles and precedents. The problem was completely solved by the mathematician Robert Aumann and Michael Maschler in the 1980s at the Hebrew University of Jerusalem [6, 18, 72].

[3] Descartes' *Cogito ergo sum* or "I Think therefore I am" is connected to the idea of rational decision making central to Game Theory: an individual's choice based on own understanding and assessment of a situation.

[4] Evolution á la Darwin has been combined with Game Theory to explore and understand how organisms evolve and adapt.

[5] Sun Tzu, circa 544–496 BCE is mostly known for The Art of War stating:
Knowing the other and knowing oneself, In one hundred battle no danger
Not knowing the other and knowing oneself, One victory for one loss

Most societies, past and present, have been and are fascinated by game playing and gambling, in particular, *games of chance or of strategies*.[6] (i.e., outcomes determined by chance or by skills and abilities): Dice, Cards, Roulette, Poker, Craps, Lotteries, Spin-to-win, Random Prize draws. Useful and practical examples for experiments, games and gambling have also contributed to the development of the mathematical theory of Probability. Indeed, there has been some serious scientific excitements and achievements strongly related to Games and Gambling; notably, the work of Girolamo Cardano in his *The Book on Games of Chance* around 1520, who also excelled in Chess and Backgammon and other dice games; Blaise Pascal and Pierre de Fermat have made foundational contributions to probability and gambling questions around 1654, encouraged by their compulsive gentleman gambler friend, Antoine Gombaud, a.k.a the Chevalier de Méré.[7] This has later led to the famous *Pascal's wager*, considered a game-theoretic approach to the belief in God: that is, it is "probabilistically prudent" to believe in God (see Packel [64]). More notably, Christiaan Huygens around 1657 introduced the concept of expectation in his treatise *On reasoning in Games of Chance (De Ratiociniis in ludo aleae.*

What exactly is game theory? In essence, Game Theory[8] is a mathematical framework (American Mathematical Classification code 91A) that allows analysis of strategic decision making between multiple entities called *the Players*. It has all the features of a mathematical construct: a concise set of concepts and assumptions, many fundamental theorems, and many real-life applications where human and social behaviors are mathematically analyzed. It has traditionally carries a *rationality assumption*, however, more concerned with the "pursuit of happiness" by the decision-makers (i.e., not necessarily altruists).

Game theory or Interactive decision-making was first initiated in the field of microeconomics. See *Theory of Games and Economic Behavior* by the mathematician John Von Neumann[9] and the economist Oscar Morgenstern. The book has a seismic effect on quantitative social sciences (economics, political science and psychology), providing the long-awaited mathematical equivalency of mathematical physics.

Not knowing the other and not knowing oneself, In every battle certain defeat
(S. Tzu in the Denma Translation, Shambala Library 2002).

[6] Playing cards appeared in Europe around the tenth century. A precursor of the six-faced die, called *Astragalus*, was found on Sumerian, Assyrian and Egyptian archaeological sites, circa 3600 BCE. Lotteries are from the first century Roman Empire.

[7] Self-knighted Chevalier de Méré, he is a typical french *salon theorist* known for his essays "L'honnête homme" and "Discours de la vraie honnêteté".

[8] Game Theory has been popularized by various means and is the theme of the 2001 Best Picture and Best Director, *Beautiful Mind*, featuring Russell Crowe playing John Nash, the nerdy Princeton student who's, alledgedly, life and dating experience led him to the Nash Equilibrium solution concept that ultimately ended in a 1994 Noble Price Award in Economics [56].

[9] Von Neumann's 1928 paper founded the theory of two-person zero-sum games, making him the mathematician most closely related to the creation of the theory of games. However, Emile Borel preceded him in formulating a theory of games.

Interactive strategic decision-making schemes prevail across most disciplines, leading to the expansion of the applicability of game-theoretic methodology over the years to areas such as Stock Markets and Financial Management, Military War Games, National Security and Anti-terrorism, Bridge games, Hunting Party, Politics, Evolutionary Biology, Computer Science. The *actions/strategies* of the *players* should result in the *best possible* consequences or outcomes according to their preferences [7, 14, 16].

First, let us agree on what a decision-making is. A good related reading is found in the book "Game Theory" by James Webb [82].

Optimum and Rational Decision Making

In a game play, *Decision making* actually refers to making optimum decision-making. That is, from an available set of courses of action or possible behaviors one first determines the outcome of each one of these actions, and then chooses the *preferred* one, in the sense of the one that maximizes a personal utility or payoff, numerical or otherwise (e.g. something of great value to one, be it monetary, social norms, etc.). Maximizing/minimizing easily translates in a mathematical framework (extremum of a function in basic calculus). Therefore, decision-making sets in motion a course of action that leads to an outcome, which is certain in the absence of randomness. A payoff function is then determined to associate a numerical value with each action. The decision-maker strives to maximize/optimize such a payoff. Note that a payoff is "invariant" under certain changes, the so-called *affine transformation or rule* assigning the same value of a payoff $\pi(x)$ to its affine transformed payoff $a\pi(x) + b$. In general, payoffs model individual preferences during the social interactions. The so-called *Darwinian fitness* in Evolutionary game models.

Uncertainty is an inherent part of any decision-making. We then talk about expected outcome/expected payoff, to be represented as a random variable and its associated probability distribution. For example, decision-making with respect to the performance of the stock market. An uncertain outcome is also called *a lottery*, which implies a finite probability distribution over the set of all payoffs. Of course, randomization is not required to maximize one's payoff; but it can serve as a tie-breaker as seen below in the case of multiple acceptable outcomes or equilibria. We will refer to a plan of action as a *strategy*, *pure* in the absence of randomization, and *mixed* otherwise.

1.2 Variation on the Nature of Games

A player could be any decision-making entity such as an automaton, a machine, a program, a person or an animal, a living cell or a molecule. The term *Game* refers to in any interactive situation leading to the sharing of clearly quantifiable benefits, the *payoffs/utilities* and costs. Traditionally, a fundamental premise has been the assumed *rationality* of all players. The strategies may be tightly coupled, strongly correlated (i.e., *entangled* as in quantum strategies), allow *superposition* (e.g., quantum game) or may allow a probability distribution for a random selection (*mixed strategies*). Strategies termed *pure* refer to simple and consistent action plans specific to a given player (acting with certainty) during a game (e.g., playing *rock* in the popular *rock-paper-scissors* game). The continuity of probabilities ensure an infinitely many mixed strategies for the player. A *totally mixed strategy* implies the assignment of strictly positive probability to every pure strategy (also called a *degenerate* mixed strategy with probability one).

Along with the rationality assumption for every player, goes also the *selfishness* assumption. That is, players strive to maximize their self-interest, i.e., the individual payoffs during a game. Preferences in game theory may include altruistic motivations, moral principles, social constraints. Oftentimes, any perceived cooperative behavior is transient and driven by selfish objectives. However, we distinguish such game called a noncooperative/competitive game from cooperative or coalitional game. In the latter game, players form coalitions or groups, with sometimes external enforcement, and focusing on, e.g., surplus or profit-sharing among the coalitions.

Competitive games may also describe the so-called zero-sum games in which the sum of the payoffs of all players is zero regardless of their actions/moves/strategies: in its two-person version one player's loss corresponds to another player's gain. A so-called payoff matrix is often used to represent a two-player zero-sum game. In solving such a game, one seeks a set of strategies to minimize the maximum loss, the *Minimax Theorem* or, inversely to maximize the minimum payoff, the *Maximin Theorem*. The solution or outcome is called a *pure saddle point*. The antithesis of such games is the *common interest game*: players have perfectly aligned interests. A particular class of zero-sum games or win-loss game scenarios is that of Combinatorial Games.[10] in which players alternate in moving. This class of games is also benefiting from the blooming development of artificial intelligence and deep neural networks. It is typically a two-player game with no hidden information (i.e., full disclosure of the game's position) and no chance elements (i.e., the next position of the game is totally determined by a player's move). There are in general two winning conditions along the "divide and conquer" approach: "Last move wins" called the *Normal play* or "last move loses" called the *Misère play*. See *Winning Ways for your Mathematical Plays*, Berlekamp et al. [12]. Combinatorial Game is rooted in ancient board games (e.g., Go, Chess, Checkers) and in the so-called *recreational mathematics* such as *The Tower of Hanoi*.[11] Indeed, mathematically the Tower of Hanoi corresponds to a *Hamiltonian Circuit* on the n-dimensional cube. Other well-known recreational mathematics include the *Chinese Rings*, aka *Baguenaudier, Cardan's*

---

[10] Initiated around 1901 by Charles Leonard Bouton, the theory of combinatorial game was further developed by Roland Percival Sprague and Patrick Michael Grundy in the 1930s, continued with John Milnor and Olof Hanner in the 1950s.

[11] Francois-Edouard Anatole Lucas, 1842–1891, is seen as the greatest French recreational mathematician and invented the Tower of Hanoi game in 1883 and whose original is the Conservatoire National des Arts et Métiers-Musée National des Techniques. n discs require $2^n - 1$ moves to solve the problem.

suspension, Cardano's rings, Devil's needle, five pillars puzzle and the Chinese *I-Ching Hexagrams*.[12] Well-known Combinatorial Games include Chess, Checkers, Tic-Tac-Toe, Go and Connect Four. Well-known games but not considered combinatorial are Bridge, Backgammon, Poker, Snakes and Ladders.

Some competitive games such as Chess may feature *perfect information* to refer to the fact that each player is perfectly aware of the previous actions of all other players, including of course the initial state of the game in a sequential game (with predefined order). On the other hand the game may be an *imperfect information* one, with some players not accessing the entirety of other players' actions.

In classical game theory, there are typically three dominant mathematical forms (1) Normal Form or Strategic Form Game (SFG) (2) Extensive Form Game (EFG) (3) Coalition Form Game (CFG) dealing with options for subsets of players. The SFG consists of players making the relevant decisions with the strategies available, the payoffs are the rewards contingent upon the actions of all the players in the game. The EFG places the emphasis on the timing of the decisions to be made, as well as the information available, and is representable by a *decision or game tree*. See more details in Fudenberg et al. [32], Webb [82].

Normal or Standard Form Game

The most common form of classical games is the so-called Normal Form/Standard Form: all players make decision simultaneously without knowledge of other players' actions. A player controls only their own actions, and is interested only in action profiles that maximize their own payoff/utility.

A formal definition or representation of a *Normal Form* game is as follows, using the terms *action* and *strategy* interchangeably, as well as *utility* and *payoff*.

Definition 1.1 A *Strategic Normal Form Game* is described by a triplet $\mathcal{G} = \langle \mathbf{P}, (\mathbf{S}_i)_{i\in\mathbf{I}_n}, (u_i)_{i\in\mathbf{I}_n} \rangle$, where $\mathbf{I}_n = \{1, \dots, n\}$, and

1. $\mathbf{P}$ is a finite set of the n players $P_{i=1,\dots,n}$.
2. $\mathbf{S}_i$ is the non-empty set of *pure strategies* or available actions for player P_i. The set S_i can be written as $S_i = \{s_1^1, s_i^2, \dots, s_i^\alpha, \dots, s_i^{n_i}\}$, where the s_i^α are the *pure strategies* player p_i has access to, and n_i is the finite number of these pure strategies. (Traditionally the payoff functions are real-valued.)
3. $u_i : \mathbb{S} = \prod \mathbf{S}_{i\in\mathbf{I}_n} \to \mathbb{F}$ is the *utility* or *payoff* function for player P_i. Thus a tuple of payoff functions $u = (u_1, \dots, u_n)$. For $u = (u_1, \dots, u_n)$ the set of payoffs is given by $u(S) = (u_i(S))_{i=1,\dots,n}$.
4. The vector $s = (s_1, \dots, s_n) \in \mathbb{S}$ denotes the *pure strategy/ action profile* or *outcome*.

[12] Book of Changes with 64 hexagrams corresponding to nature fundamental forces: the Yin for the 6 broken lines and the Yang for the 6 unbroken ones.

5. We denote by $s_{-i} = (s_1, \ldots, s_{i-1}, s_{i+1}, \ldots, s_n)$ the vector of strategies taken by all other players except player P_i. That is, $s = (s_i, s_{-i})$. And accordingly $\mathbb{S}_{-i} = \prod_{j \neq i} \mathbf{S}_j$ describes all action profiles for all players except for P_i.

A "play" of the game refers to the n-tuple strategy profile or pure state

$$G := s^{\alpha} = (s_1^{\alpha_1}, \ldots, s_n^{\alpha_n})$$

where player P_i chooses to play the strategy $s_i^{\alpha_i} \in S_i$, and received a reward or payoff computed by the payoff function

$$u_i(s^{\alpha}) = u_i(s_1^{\alpha_1}, \ldots, s_n^{\alpha_n})$$

The reward or payoff, in general, depends on the strategies played by all the players, under the rational assumption that each tries to maximize their payoff. Thus a need of a compromise or an *equilibrium* or *solution* of the game.

For games of "complete information" players know the strategy and payoff spaces of each other.

Remark 1.2 Note that, classically, the utility or payoff function is real-valued. The co-domain $\mathbb{R}$ is endowed with the Euclidean/Archimedean norm/absolute value $|.|_{\infty}$. A recent advance, described below, is to consider p-adic/non-Archimedean valued payoff function to more effectively address nuances and the possible hierarchy in the payoffs structures. That is, the co-domain is the payoff function u_i is $\mathbb{Q}_p$, the set of p-adic numbers defined below, endowed with p-adic norm $|.|_p$. See Gouvêa [35].

Definition 1.3 (*Dominant Strategy*) A pure strategy/action or value $\alpha \in S_i$ is said to be *(strictly) dominated* by a strategy $\beta \in S_i, \forall i = 1, \ldots, n$ denoted by $\alpha < \beta$ when $u_i(\alpha, s_{-i}) < u_i(\beta, s_{-i})$.

Remark 1.4 Dominated strategies/actions/profiles are *rationally unjustifiable.* However removing them in a given game may lead to the emergence of other dominated strategies in the subgame, which results inductively in the so-called *iteratively dominated strategies. Iteratively undominated strategies* survive all rounds of elimination by dominance, resulting sometimes in instances of games called *dominance solvable* where the undominated strategies form a singleton (see Fisher [31], Fudenberg and Tirole [32]).

Solution Concepts

Most of the time games cannot be solved just by removing dominated strategies. The *solution concept or equilibrium* refers to a balance of players' strategies in such a way that no player has a motive to change unilaterally. The equilibrium is *strong* if it does not allow any group deviation, and it is *Pareto efficient or optimal*[13] if there is no other outcome to make all players better off. The rationality assumption indeed forces the appearance of an equilibrium solution.

We then refer to the fundamental solution concept that is needed to capture the steady state of play with each player acting optimally given a "conjecture" about the behavior of other players. It is the *Nash Equilibrium* of the game to characterize strategy profiles resilient to unilateral deviations. It is akin to a compromise in a game where every player endeavors to maximize their own selfish payoff. Formally

Definition 1.5 A strategy profile $s^* = (s_1^*, \ldots, s_n^*) = (s_i^*, s_{-i}^*) \in \mathbb{S}$ is a Nash Equilibrium (NE) if

$$u_i(s^*) = u_i(s_i^*, s_{-i}^*) \geq u_i(s_i, s*_{-i}), \quad \forall s_i \in S_i. \tag{1.1}$$

Remark 1.6

- The *Nash Equilibrium* must be *strategically stable* to unilateral deviations: No player can unilaterally profitably deviate. It is *strong* if it is stable to group deviations, as to mitigate mutation and the impact of mutant players.
- The *Nash Equilibrium* exists for a broad class of games as proved by the Nash's famous theorem below. See Nash [57, 59, 60].
- Some instances of games may be described by admissible mechanisms for coordination to ensure emergence of desirable collective behavior with respect to a given objective, thus leading to a Nash Equilibrium.
  For example, consider the simple game of the *commuting time to get home from work* which depends not only on the chosen route but also on decisions taken by other drivers. A solution concept or a *Nash Equilibrium (NE)* is: *In traffic everyone is driving on the right. No single driver (rational) has an interest in driving on the left.*
- A NE may result from repeated interactions, even in a game with partial information and primitive decision-making as in the *Harper's 33 Ducks Experiment* we here recall [37].
  In the botanic garden of Cambridge University, UK, the biologist David harper experimented with a flock of 33 ducks and 2 bread tossers, 20 m apart feeding the ducks at regular intervals. Tosser 1 (the least profitable site) has a frequency of supply of 12 pieces of bread per minute whereas Tosser 2 (the most profitable site) has a frequency of 24 pieces per minute. After about a minute, the number of ducks at the least profitable site stabilizes around 11, with thus 22 ducks at the most profitable site, which indicates a Nash equilibrium has been reached: Switching unilaterally to another site for a duck would result in getting less food. Indeed the amount of food a duck gets does not depend on its choice but also on others' choice. During a transient period, behaving as a typical optimizer most of the ducks rushed to the most profitable site, for some to realize they could get more food at the least

[13] Vilfredo Pareto, Italian mathematician, defines the efficiency as "A situation where there is no way to rearrange things to make at least one person better off without making anyone else worse off".

profitable site, and therefore switched site. After switching a few times the ducks stick to a given choice. Repeated interactions has led to a Nash Equilibrium in a game where the entities, the duck, could hardly qualified as rational players. Such a process has been termed "*tâtonnement*". (Iterative adjustments in strategies until equilibrium is reached) [21].

Remark 1.7 (*Historical Comments*)

- An intuitive explanation of the concept of *Nash equilibrium* could be traced back to Antoine Augustin Cournot in his *Recherches into the Principles of the Theory of Wealth in 1838* in which one could find as well an evolutionary or dynamic idea of the best response correspondence. From Cournot's works Francis Ysidro Edgeworth in his *Mathematical Psychics* derived the concept of competitive equilibria in a two-players economy. Then Emile Borel, in *Algèbre et calcul des probabilités* published in Comptes Rendus de l'Académie des Sciences, vol. 185 (1927), dealt with mixed strategies, probability distributions leading to stable game.
- However the modern analysis, a milestone in the history of game theory, is accepted to have been initiated by John von Neumann and Oskar Morgenstern [81] in their book *Theory of Games and Economic Behavior*, drawing from which John Nash provides us with the modern methodological framework. Von Neumann proved there his famous Minimax Theorem for zero-sum games.
  Other seminal works include that of Nash on Nash Equilibrium [59] and on Nash Barganining [58], and Shapley on the Shapley Value and games with transferable utility [74]. The work by Luce and Raiffa [53] is by now classical, as many examples traced back to this source, such as *Prisoners' Dilemma* and the *Battle of the Sexes*. However one of the first formal treatise on game theory is by Zermelo [86], a logician who proved that in the game of Chess either White always wins or Black always wins or either a player can always force a draw.
- Classical applications of game theory were in Economics, Behavioral sciences and Biology. Game Theory and its impact on the theory of Economics has led to several Nobel Prize winnings, starting with the 1072 Nobel Prize to Ken Arrow; followed by the 1994 to John Nash, Reinhard Selter, John Harsanyi; then Mirrlees and Vickrey (1996), Sen (1998), Akerlof et al. (2001), Aumann and Schelling (2005), Hurwicz et al. (2007), Roth and Shapley (2012). As a consequence, ideas, concepts and formal language of game theory form large parts of economics.
- The most recent applications are related to networked systems as reflected in Online advertisement on the Internet, Information evolution, Belief propagation in social networks; deployment of distributed passive and active sonars for underwater sensor field design. Game can be used indeed to model the interactions taking place in a network, with the network nodes, largely interdependent, acting as players competing or forming coalition to maximize their *quality of service*. For instance in the field of Signal Processing for communication networks one may design a game to address issues such as data security, spectrum sensing in cognitive radio, multimedia resource management and image segmentation.

Existence and Computation of Equilibrium

Nash [59] famously proves that Every finite, noncooperative strategic game of two or more players has at least a (possibly mixed) Nash Equilibrium.

However, this is only an existence result; that is, it does not present the steps to find an equilibrium. The best response correspondence is one way for finding a NE.

For any $s_{-i} \in S_{-i}$ define the *best response correspondence* of player P_i as

$$B_i(s_{-i} := \{s_i \in S_i \quad | \quad u_i(s_i, s_{-i}) \geq u_i(s_i', s_{-i}) \quad \forall s_i' \in S_i\} \tag{1.2}$$

A strategy profile s^* is a NE if

$$s_i^* \in B_i(s_{-i}) \quad \forall i = 1, \ldots, n.$$

In other words, the Nash Equilibria are the fixed points of best-correspondence mapping

$$B(s) := (B_i(s_{-i})_{i=1,\ldots,n}.$$

Mixed Strategy Nash Equilibrium

Let consider a probability distribution $\Delta(S_i)$ over the pure strategy set S_i. We call $\sigma_i \in \Delta(S_i)$ the mixed strategy of player P_i, and $\sigma = (\sigma_i)_{i=1,\ldots,n}$ is the mixed strategy profile. That is, $\sigma_(s_i)$ is the probability assigned to each strategy s_i.

The expected payoff/utility is defined as

$$\tilde{u}_i(\sigma_i, \sigma_{-i}) := \mathbf{E}(u_i) = \sum_{s \in S} \left(\prod_{i=1,\ldots,n} \sigma_i(s_i) \right) u_i(s) \tag{1.3}$$

Therefore

Definition 1.8 (*Mixed NE*) A mixed strategy profile σ^* is a Mixed Nash Equilibrium (MNE) if

$$\tilde{u}_i(\sigma^*) \geq \tilde{u}_i(\sigma_i, \sigma_{-i}^*) \quad \forall i = 1, \ldots, n \quad \forall \sigma_i \in \Delta(S_i) \tag{1.4}$$

Remark 1.9 (*Pareto-Optimality*)

The NE or any equilibrium is said to be *Pareto-inefficient* whenever there is another strategy profile in the game providing a payoff greater than the equilibrium payoff. This constitutes a drawback for considering a NE.

A strategy profile s^* is therefore considered *Pareto-Optimum* if there is no other strategy profile s such that $u_i(s) \geq u_i(s^*) \quad \forall i = 1, \ldots, n$ and $u_{i_0}(s) > u_{i_0}(s^*)$ for some i_0.

That is to say that at a Pareto-Optimum NE, the payoff to one player cannot be increased without decreasing the payoff to at least one other player. A classic example of an inefficient NE is in the well-known *Prisoner's Dilemma*: The unique NE is *Pareto inferior*: indeed, if both players choose (cooperate, cooperate), they would both get a higher payoff.

In addition to these characteristics an equilibrium solution may well feature other socially sounding characteristics, such as a *maximizer of social welfare*, originated in the application of game theory to economics [23]. Indeed

Definition 1.10 A strategy profile s^* is called social optimum if it satisfies $s^* \in \arg\max_{s\in S} \sum_{i=1,\dots,n} u_i(s)$

As one of the most important solution concepts, the efficiency of the Nash Equilibrium has to be constantly assessed. It can also be measured using the following concept of Price of Anarchy (PoA) defined as

$$PoA := \frac{\max_{s\in S} \sum_{i=1,\dots,n} u_i(s)}{\min_{s\in S^*} \sum_{i=1,\dots,n} u_i(s)}, \tag{1.5}$$

where S^* is the set of all Nash Equilibria in the game.

That is to say that the *Price of Anarchy* measures the performance loss of the worse NE when compared to a socially optimum strategy. The efficiency of the NE is higher whenever the PoA is closer to 1.

Mechanism Design refers to a process of improving the NE efficiency by transforming the game while keeping the NE, noting, for instance, that payoffs are invariant under *affine transformation*. Of course, one may also improve efficiency by modifying the solution concept with the game unchanged, Such modification may refer to equilibrium concept such as Correlated Equilibrium (CE) or the Nash Bargaining Solution(NBS), both here recalled, See also Nash [59] and Aumann [4] in the references.

Definition 1.11 (*Correlated Equilibrium*)
A *Correlated Equilibrium* is a joint probability distribution $\sigma^* \in \Delta(S)$ satisfying

$$\sum_{s_{-i}\in S_{-i}} \sigma^*(s_i, s_i)u_i(s_i, s_{-i}) \geq \sum_{s_{-i}\in S_{-i}} \sigma^*(s_i', s_i)u_i(s_i', s_{-i}), \tag{1.6}$$

where $i = 1, \dots, n$ and $s_i' \in S_i$.

See more details in Aumann's work [4]: A Traffic light with its 3 distinct alternating colors of red, yellow and green, acts as a trusted neutral third party coordinating device, is considered as a simple example of correlated equilibrium: no driver has an incentive to disobey a traffic light while everyone else is following it. It serves as a joint probability distribution correlating the actions of all drivers, with a high probability of one driver seeing green while the other driver sees red.

The CE is said to be *Coarse* if it requires an a-priori commitment to follow the given recommendation, e.g., getting a driving license implies an a-priori commitment to obey a traffic light while driving. We indicate below how *entanglement* of quantum strategies could be considered strong non-classical correlation between strategies.

Seeking an alternate efficient may also lead to the so-called *Nash Bargaining Solution* (NBS) defined here for a two-player game as

Definition 1.12 (*Nash Bargaining Solution*) (NBS)

The Nash Bargaining Solution is the unique Pareto-Optimal solution to

$$\max_{(u_1,u_2)\in\mathscr{U}\,|} (u_1 - \lambda_1)(u_2 - \lambda_2) \tag{1.7}$$

$$\text{subject to} \quad u_1 \geq \lambda_1, \quad u_2 \geq \lambda_2, \tag{1.8}$$

where $\mathscr{U}$ is a closed convex set of utility points. The product $(u_1 - \lambda_1)(u_2 - \lambda_2)$ is called the *Nash product.*

Potential and Congestion Games

Many game scenarios are concerned with finding a *pure* Nash equilibrium. There are two well-known classes of games which always possess at least one pure Nash equilibrium. We recall the following result in the case of infinite strategy set games.

Theorem 1.13 (Debreu, Fan, Glicksberg) *Consider a strategic form game with an infinite strategy set such that*

1. *S_i is convex and compact*
2. *$u_i(s_i, s_{-i})$ is continuous in s_{-i}, continuous and quasiconcave in s_i.*

Then the game has a pure Nash Equilibrium.

See proof and more details in Debreu, Fan or Glickberg 1952.

Definition 1.14 (*Potential Games*)

A game is a *potential game* if there exists a potential function Φ, real-valued over the strategy set $\mathbb{S}$ such that for every player P_i and for every strategy profile $s = (s_i, s_{-i})$ and every strategy s_i' we have

$$\Phi(s_i, s_{-i}) - \Phi(s_i', s_{-i}) = u_i(s_i, s_{-i}) - (s_i', s_{-i}) \tag{1.9}$$

That is, in potential games the change in the player's utility upon switching of strategy remains the same as the change in the potential function [66].

It was then proved that

Theorem 1.15 *Every potential game has at least one pure Nash equilibrium, namely the strategy that maximizes the potential function Φ.*

Here also there is no indication on how to determine that pure equilibrium. However, the proof of this theorem reveals that the *best response dynamics* leads to finding the pure Nash equilibrium, without saying much about the number of steps. That is

Theorem 1.16 *In any finite potential game, best-response dynamics always converge to a pure Nash equilibrium.*

Definition 1.17 (*Congestion Games*)
A *congestion game* is defined by a finite set P of n players, a finite set S of resources. Each player P_i has a set $S_i \subset S$ of allowable strategies. For each resource $r_i \in S$ and assuming x_j players using r_j, there is a load dependent cost function $c_j(x_j)$ so that the total cost to a player P_i is $\sum_{j \in S_i} c_j(x_j)$.

The simplest example of congestion game is the non-atomic network routing game. It was proved that

Theorem 1.18 ([66]) *Every congestion game is a potential game*

The implication is, therefore, that every congestion game has a pure Nash found using best response dynamics.

Mean Field Game

The study of decision-making in large population of interacting agents is much more involved than in the above scenarios, and it is the subject of the sub-game theory of Mean Field Game (MFG), introduced by Jean-Michel Lasry and Pierre-Louis Lions and others in the 2000s [52]. Its focus is on the collective or average behavior of the game population rather than the individual one.

The *Hamilton-Jacobi-Bellam equation* and the *Fokker-Planck equations* are some of the most common equations studied in MFG, respectively addressing the optimal control of every single player and the evolution over time of the distribution of players [25]. These fundamental equations of MFG have led to some interesting insights on existence and uniqueness conditions for equilibria. They are respectively

$$-\partial_t u(x,t) + H(x, \nabla u) = f(x,m) \quad \text{(Hamilton-Jacobi-Bellam)}$$

where $u(x,t)$ is the value function, H the Hamiltonian and $f(x,m)$ the cost function in terms of population density $m(x,t)$. It is considered a "backward PDE".

$$\partial_t m - \nabla \times (m \nabla H_p(x, \nabla u)) = 0, \quad \text{(Fokker-Planck)}$$

with H_p the gradient of the Hamiltonian with respect to momentum accounting for the drift in the players dynamics. It is considered a "forward PDE" on the population distribution.

Solutions methods are varied, and oftentimes ad hoc, and include: fixed-point techniques. Variational schemes (convex optimization), and of course, Numerical schemes (Finite Difference, Finite Elements, neural-network-based solvers).

The following Fixed Point Theorems are widely used in MFG:

Theorem 1.19 (Schauder Fixed Point Theorem) [71]

If $\mathbb{X}$ is a Banach space and $T : \mathbb{X} \to \mathbb{X}$ is a compact, continuous operator mapping a convex, closed, and bounded subset of $\mathbb{X}$ into itself, then T has a fixed point.

and also

Theorem 1.20 (Banach Fixed Point Theorem) [9]

If $(\mathbb{X}, d)$ is a complete metric space and $T : \mathbb{X} \to \mathbb{X}$ is a contraction mapping (i.e., there exists $0 \leq c < 1$ such that $d(T(x), T(y)) \leq cd(x, y)$ for all $x, y \in \mathbb{X}$), then T has a unique fixed point x^* such that $T(x^*) = x^*$.

In some notable recent developments, MFG provides an ideal framework for modeling financial markets and resource allocation, traffic flow and communication networks, epidemics, and multi-agent reinforcement learning and swarm intelligence [20].

MFG has been experiencin growing relevance in analyzing and proposing equilibrium solutions for complex systems with large numbers of interactive decision-makers, leveraging powerful computational algorithms. It has been extended to the so-called Ergodic MFG to investigate long-term behavior in large population interactive decision-making, with time-independent solution equilibrium.

Approximate Nash Equilibrium

Recall that Nash's famous theorem is only an existence theorem about Nash Equilibrium (NE), based on Brouwer and Kakutani's fixed point theorems, respectively recalled as

Theorem 1.21 (Kakutani Fixed Point Theorem) [44]

Let $\mathbb{K} \subset \mathbb{R}^n$ be a compact and convex subset and $F : \mathbb{K} \to 2^{\mathbb{K}}$ a mapping into the set of compact and convex subsets of $\mathbb{K}$ which is upper semi-continuous[14] *in every $x \in \mathbb{K}$, then $\mathbb{F}$ possesses a fixed point, i.e., there is some $\hat{x} \in \mathbb{K}$ with $\hat{x} \in F(\hat{x})$.*

It is a generalization of the Brouwer Fixed Point Theorem stating:

Theorem 1.22 ([19])

Every continuous mapping of a convex and compact subset of an n-dimensional Euclidean space into itself.

The computation of the NE remains challenging and complex even with the advances in computational methods assisted by AI/ML. To date no efficient algorithm is known. Known algorithms include:

- *Regret Minimization*[15] (no-regret learning): strategies are adjusted taking into account players past performance to minimize regret

---

[14] F is upper semi-continuous in $x \in \mathbb{K}$, if $\forall (x_k)_{k\in\mathbb{N}} \in \mathbb{K}\ \ x_k \to x$ and $\forall (y_k)_{k\in\mathbb{N}} \in \mathbb{K},\ \ y_k \in F(x_k),\ \ y_k \to y$ then $y \in F(x)$.

[15] Popularized by Jeff Bezos after leaving his well-paid hedge fund job to start Amazon, as he might have regretted not pursuing his entrepreneurial vision.

- *Reinforcement Learning*: Decision-making through interactions with environment to include merit-based rewards/penalties to maximize cumulative rewards over time.
- *Linear Programming*, in particular, for zero-sum games, in an attempt to compute near-optimal strategies: determine optimal solution to problems with linear relationships, constraints and objective function (e.g., maximize profit and minimize cost).

As it is often the case in mathematics, one way around the problem has been the notion of *Approximate Nash Equilibrium* (ANE) to some ϵ accuracy. That is, for a game in a pure or mixed normal form, one would like to evaluate a minimal tolerance value to express how far a strategy profile $\hat{s}$ is from the exact NE s^* whose existence is guaranteed in the Nash Existence Theorem. In other words,

Definition 1.23 A strategy profile $\hat{s} = (\hat{s}_i, \hat{s}_{=i})$ is an ϵ-Nash Equilibrium ($\epsilon \geq 0$) if no unilateral deviation for the exact NE s^* leads to an expected payoff gain of more that ϵ. Mathematically,

$$u_i(\hat{s}) \geq u_i(s_i, \hat{s}_{-i}) - \epsilon, \quad \forall i = 1, \ldots, n \quad \forall s_i \in S_i \tag{1.10}$$

Due to the intractability of computing Nash equilibria, approximations seem more practical and feasible, as it is in general in numerical analysis or approximation theory. Moreover, in real-world scenarios, games of chance or games of strategy, agents/players act optimally within a "reasonable" margin error, despise the "rationality" assumption.

We now turn to one of the most important class of games.

2 Evolutionary Game Theory

2.1 Introduction and Tenets of EGT

Evolutionary Game Theory (EGT)[16] was adapted from the biological evolution theory and developed the mathematical theory of Darwinian evolution, providing the most promising applications of the theory of games: it has a profound impact on social and behavioral sciences from economics to psychology to linguistics and language evolution to learning and cooperation in agent-based systems (origin of social norms and cultural trends). EGT contributes to addressing questions such as whether a fly, a fig tree or the Harper experiment ducks, all be decision-makers evaluating all possible outcomes to select their optimal strategy for "success".

[16] EGT first appeared in the work "The Logic of Animal Conflict" of Maynard Smith, a mathematical biologist and George Smith, attempting to understand ritualized animal behavior.

From the early works of Smith and Price statements, the subsequent theory of evolution was based on *games of selfish genes* rather than considering organisms acting *for the good of the species* [76].

We complete this introduction of EGT with some quotes from Malthus and Darwin T. R. Malthus: *Through the animal and vegetable kingdom, nature has scattered the seeds of life abroad with the most profuse and liberal hand; but has been comparatively sparing in the room and nourishment necessary to rear them.*

To which, Darwin reacted: *Fifteen months after I had begun my systematic enquiry, I happened to read for amusement "Malthus on Population"... it at once struck me that ... favorable variation would tend to be preserved and unfavorable ones to be destroyed. Here, then, I had at last got a theory by which to work.*

Remark 2.1 EGT mainly considers a large population of individuals, the players in the broader sense. These players P_i have a finite number N of strategies $s_{i=1,\dots,n}$ each of which has a frequency or a probability of occurrence that changes over time in response to the decisions of all the other players under the principle that individuals switch to those strategies with better payoffs from poorly performing strategies.

The changes in frequency of the strategies in the population are described by a probability distribution $\Delta(S)$ on the finite set $S = (s_{i=1,\dots,n})$ of the strategies.

2.2 *Evolution Matrix Game*

Consider a large population whose individuals have n strategies $s_i \in S$ for interaction with each other, together with a probability distribution $\Delta(S)$ over the finite set of strategies. That is, let $p_i \in [0, 1] \subset \mathbf{R}$, be the probability for strategic value s_i to be chosen. The vector of probabilities or frequencies $p = (p_1, \dots, s_n) \in [0, 1]^n \subset \mathbf{R}^n$ describes a state of the population as well as a behavioral strategy profile , with the components s_i satisfying the probability constraint $\sum_i p_i = 1$. The state vectors $p = (p_{i=1,\dots,n}$ lies on the (n-1) simplex space of strategies

$$\Delta = \{s = (p_1, \dots, p_n) \in \mathbf{R}^n \ \mid \ 0 \leq p_i \leq 1, \ \sum_i p_i = 1\} \tag{2.1}$$

compact and convex spanned by the set of vectors e_i of the standard basis of the Euclidean space $\mathbf{R}^n$.

$$e_i = (0, \dots, 0, 1_i, 0, \dots, 0), \quad i = 1, \dots, n. \tag{2.2}$$

Remark 2.2

- Pure populations states are given by the corners vectors of the standard Euclidean basis $e_i = (0, \dots, 0, 1_i, 0, \dots, 0)$, $i = 1, \dots, n$. These states are also called *homogeneous*.

- Note that the probability p_i assigned to strategy s_i changes with respect to payoffs by learning, copying, inheriting etc. And in turn payoffs are function of the probabilities yielding a *feedback loop dynamic*. The difference $h_i(s)$ has been identified in certain "game dynamics" with a relative increase of the frequency.
- Strategies with higher payoffs reproduce faster. Consequently individuals have a tendency to vary the probability assigned to a core value based on some perceived relative advantage.
- Subscribing to competing strategic values lead to the emergence of mutant subgroups, with the by-products of the appearance of a variety of life forms, beliefs, cultures, languages, practices or techniques.
- In contrast to other types of dynamics such as Replicator, Replicator-Mutator, or Logit Dynamics, any kind of revision protocol is described by changes in the values of the probability assigned to each strategy, not by switching of strategy. Successful strategies are represented by "successful individuals" and could be learned (imitated), possibly inherited [40–42].

The interaction between an s_i-individual and an s_j-individual results in a *payoff/reward/utility* denoted by $a_{ij} \in \mathbf{R}$. Therefore all the payoffs form the *evolution payoff matrix*

$$A = (a_{ij})_{1 \leq i,j \leq n} \tag{2.3}$$

defining the game as an evolution matrix game. Based on the context, the real entries of this $n \times n$ matrix could be assumed to satisfy various properties of matrices. That is, the matrix A could be *symmetric*, i.e., $A = A^T$; *skew-symmetric*, $A = -A^T$; *cyclic symmetric* (Toeplitz circulant with indices counted cyclically modulo n); or a *banded matrix* (sparse with non-zero entries confined to a diagonal band). These features may help to further analyze the matrix.

Accordingly we define a payoff or utility function $u = (u_1, \ldots, u_n) : \Delta \subset [0, 1]^n \to \mathbf{R}^n$, also called the *payoff vector field*.

Definition 2.3

1. Given a population state determined by the probability vector $p = (p_1, \ldots, p_n) \in \Delta$, the *expected payoff* of an s_i-player is defined as

$$u_i(p) = e_i A p^T = \sum_{j=1}^{n} a_{ij} p_j = (Ap)_i. \tag{2.4}$$

   The payoff functions u_i are continuous linear functions.
2. The *average payoff/utility* in the population state p is given by

$$\bar{u}(p) = pAp^T = \sum_{1 \leq i,j \leq n} p_i A p_j. \tag{2.5}$$

3. We denote by
$$h_i(p) = u_i(p) - \bar{u}(p), \tag{2.6}$$
that is, the excess payoff between the individual expected payoff and the average payoff in the population state, which actually impacts the frequency and the probability value of a given strategy.

Note that the sign of excess payoff $h_i(p)$ for the strategy s_i in the population state p determine its variation in frequency as well as its corresponding probability. In the *standard or regular* population state p each strategy s_i receives a nonzero probability, that is, $p_i \in (0, 1)$. We denote for a general n the support of a state p by

$$supp(p) = \{i \in \{1, \ldots, n\} | \quad p_i > 0\}$$

We now define the equilibrium of such an evolutionary game.

Definition 2.4 A population state $p^* = (p_1^*, \ldots, p_n^*) \in \Delta$ is called a *Nash Equilibrium*, denoted NE, if

$$\bar{u}(p^*) = p^* A p^{*T} \geq p A p^{*T}, \quad \forall p \in \Delta. \tag{2.7}$$

Equivalently we have

$$\bar{u}(p^*) = p^* A p^* = max_{i=1,\ldots,n} u_i(p^*)$$
$$u_i(p^*) = u_i(p_i^*, p_{-i}^*) \geq u_(p_i, p_{-i}^*)$$

where again

$$u_i(p_i, p_{-i}^*) = u_i(p_1^*, \ldots, p_{i-1}^*, p_i, p_{i+1}^*, \ldots, p_n^*), \quad \forall p_i \in \Delta_i = [0, 1].$$

Remark 2.5 (*Interpretation*) In the Nash Equilibrium state p^* the unilateral change of the probability of a single strategy does not lead to a higher payoff. Under the rationality assumption of the game one would expect such a state to be maintained. That is not in general the case because individuals do not always behave rationally. Deviations are expected dictated by changes in preferences, and the stability of the Nash Equilibrium state p^* is not guaranteed. We are therefore interested in dynamic conditions that could ensure the *evolutionary stability* of that equilibrium as defined next.

Definition 2.6 (*Evolutionary Stability*) A Nash Equilibrium state $p^* \in \Delta$ is *evolutionarily stable* if

$$\bar{u}(p^*) = p^* A p^* = p_0 A p^{*T} \quad \text{for some } p_0 \in \Delta, p_0 \neq p^* \tag{2.8}$$

implies
$$\bar{u}_(p_0) = p_0 A p_0 < p^* A p_0^T. \tag{2.9}$$

That is, if the same payoff of the NE state p^* could be achieved in some other state p_0, then this other state cannot be a NE state. Conditions are needed to characterize dynamically both states, necessary and/or sufficient conditions. Here are some results widely proved in the literature on Evolutionary Game Theory. See for instance Hofbauer and Sigmund [40, 41].

- Let $p^* \in \Delta$ be a NE state with support $supp(p^*)$. Then we get

$$u_i(p^*) = \bar{u}(p^*) \quad \forall i \in supp(p^*)$$

  In a NE state, the individual expected payoff is the same as the average in the state for the nonzero probabilities making the NE state.
- As indicated above the payoff functions u_i are continuous linear functions on the $(n-1)$-simplex Δ, a convex compact subset of $\mathbf{R}^n$. This ensures $\forall p^* \in \Delta$, $\exists \tilde{p}_i \in \Delta_i = [0, 1]$, such that

$$u_i(\tilde{p}_i, p^*_{-i}) \geq u_i(p_i, p^*_{-i}) \quad | \quad \forall p_i \in \Delta_i$$

Remark 2.7 The population game as described above is so far in a *static form*. Players as rational decision-makers adapt their strategies over time by varying the probabilities assigned to each strategy for a more efficient distribution of their preferences and behaviors. Indeed strategies with higher payoff/rewards spread quickly in the population through, e.g., learning, copying/imitating. Payoffs depend on the frequencies/probabilities of strategy in the population, which themselves change according to the payoffs/rewards, resulting in a feedback loop dynamic characterizing the evolution of the population. We therefore introduce a dynamic in the game to account for the changes over time of the probability distribution $p(t) = (p_i(t))_{i=1,\ldots,n}$.

Towards the Replicator Dynamics

The per capita growth rate of the time-continuous dependent probability $p_i(t)$ is the logarithm derivative

$$\frac{d(log(p_i(t)))}{dt} = \frac{\dot{p}_i}{p_i} \tag{2.10}$$

is determined by the difference between the expected payoff to the core value A_i and the average payoff to the population state, leading to the time-continuous dynamical system or Replicator Dynamics ($\mathscr{RD}$)

$$\frac{\dot{p}_i(t)}{p_i} = h_i(p) = u_i(p) - \bar{u}(p), \tag{2.11}$$

or equivalently

$$\dot{p}_i(t) = p_i(u_i(p) - \bar{u}(p)) = p_i((Ap)_i - pAP^T) = g_i(p), \quad (2.12)$$

Remark 2.8

1. This is a system of differential equations for $i = 1, \ldots, n$ It appears in various areas such as population genetics, chemical networks, and is famously known as the *replicator dynamics*[17] introduced by Taylor and Jonker [77] and coined as such by Schuster and Sigmund [73].
2. Under the constraint $\sum_{i=1}^{n} p_i(t) = 1$ the system of equations actually reduces to $(n-1)$ differential equations we analyze on a $(n-1)$-simplex.
3. Moreover adding a constant to a column entries of the payoff matrix A does not change the equations and its dynamic properties. One may then set the diagonal entries to zeros, or set the last row of A to zero in analyzing the dynamics. Such simplifications are indeed undertaken in practice.
4. The well-defined power product $V(p) = \prod_i p_i^{\alpha_i}$ satisfies

$$\dot{V} = V \sum \alpha_i[(Ap)_i - pAp^T]. \quad (2.13)$$

5. The frequency of a pure strategy in the society as given by the probability increases when it has above average utility/payoff. Individuals have limited and localized knowledge of the whole system, according to the *distribution of information*. Some strategies or core values could become extinct as time goes to infinity. Indeed whenever a core value s_i is recursively strongly dominated as defined above, it will not survive the evolution. That is, for the assigned probability $p_i(t)$, we have $\lim_{t\to\infty} p_i(t) = 0$. The corresponding core value therefore goes extinct. This process might explain the steady variation of strategic values and norms.

Fundamentally it is shown, e.g., in Weibull [83] that the mean average payoff or utility of a population state increases along any trajectory of the evolution equation ($\mathscr{R}\mathscr{D}$) giving by the equation

$$\dot{\bar{u}} = 2 \sum_{i=1}^{n} p_i(u_i(p) - \bar{u})^2. \quad (2.14)$$

It is similar to a result by the biologist Fisher with respect to Natural Selection [31].

We analyze the system on the unit $(n-1)$-simplex subset of $\mathbf{R}^n$, and the state space of the vectors of probabilities. It is actually a cubic polynomial dynamic, whose

[17] A replicator is an entity with the means to make accurate copies of itself. It can be a gene, an organism, a belief, a technique, a convention, a cultural norm.

class has been extensively investigated. We denoted $\dot{\Delta} = int(\Delta)$ the *interior* of the simplex given by the set

$$\dot{\Delta} := int(\Delta) = \{p = (p_1, \ldots, p_n)| \quad p_i > 0 \quad \forall i = 1, \ldots, n\}. \tag{2.15}$$

The *boundary faces* are denoted by $\partial\Delta$ and defined as

$$\partial\Delta(J) = \{p \in \Delta : p_i = 0, \quad \forall i \in J\}, \tag{2.16}$$

where J is any nontrivial subset of $\{1, \ldots, n\}$. Note that the hyperplanes $\sum p_i = 1$ and $x_i = 0$ are invariant, as well as the simplex faces. See Weibull [83], Hofbauer et al. [41].

The rest points are the zeros of $g(p) = (g_i(p)) = 0$. That is, points $p \in \Delta$ such that $(Ap)_i = pAp^T$, for all $i \in supp(x)$. Thus an interior rest point is a solution of the system of linear equations

$$(Ap)_1 = (Ap)_2 = \ldots = (Ap)_n. \tag{2.17}$$

The following so-called Folk Theorem of Evolutionary Game Theory contains all relevant results of the Replicator Dynamics.

Theorem 2.9

1. *If p^* is a Nash Equilibrium then it is a rest point for $(\mathscr{RD})$.*
2. *If p^* is an evolutionarily stable NE then it is asymptotically stable for $(\mathscr{RD})$ globally if p^* is interior.*
3. *If the rest point $\hat{p}$ is stable then it is a Nash Equilibrium.*
4. *Every interior rest point $\hat{p} \in \dot{\Delta}$ is a Nash Equilibrium.*
5. *If a rest point $\hat{p}$ for $(\mathscr{RD})$ is also the forward limit point of an interior orbit $x(t)$ of $(\mathscr{RD})$, i.e., $x(t) \in \dot{\Delta} = int(\Delta)$ then $\hat{p}$ is a Nash Equilibrium.*

Remark 2.10 The converse is not true in the *Folk Theorem of Evolutionary Game Theory*.

Additional important well-known results [41] of EGT are summarized as follows.

Remark 2.11 For a boundary rest point $\hat{p}$ the difference $h_i(\hat{p}) = (A\hat{p})_i - \hat{p}A\hat{p}^T$ is also an eigenvalue $\lambda_{\hat{p}}$ for the Jacobian $J(\hat{p})$ with an eigenvector transversal to the face $\hat{p}_i = 0$. That entails A rest point $\hat{p}$ for the evolution equation $(\mathscr{RD})$ is a Nash Equilibrium if and only if its transversal eigenvalues are non-positive.

Another important question is the following: Are these strategies *permanent and resistant* to shocks of any nature and size, random or otherwise? It amounts to the notion of *permanence* or *self-sustained* for the *replicator dynamics* in the sense:

Definition 2.12 The evolution equation $(\mathscr{A})$ is *self-sustained* if $\exists$ a compact set $K \subset int(\Delta)$ such that $\forall x \in int(\Delta)$ $\exists$ a real time T such that $p(t) \in K \quad \forall t > T$.

It then results

Theorem 2.13 *If the evolution equation* ($\mathscr{RD}$) *is self-sustained then it has a unique rest* p^* *in the interior* $\dot{\Delta}$ *which is therefore the unique national Nash Equilibrium* NE_A.

Moreover along each interior orbit $p(t)$ *the time averages converge to* p^*

$$\bar{a}_i := \frac{1}{T}\int_0^T p_i(t)dt \longrightarrow p_i^*, \tag{2.18}$$

as $T \to \infty$, *and* $i = 1, \ldots, n$.

The system is said to be *ergodic*. See again Hofbauer et al. [41].

Consequently, if an interior orbit $p(t)$ has a boundary ω-limit point the time average need not converge. And every orbit converges to the boundary $\partial\Delta$ in the absence of an interior rest point.

It will be very interesting to analyze the dynamics of these subgames (i.e., using a reduced number of pure strategies) and their relation to the overall dynamics, with the possibility of appearance of special dynamics such as *limit cycles*.

Remark 2.14 Before presenting in the following sections on the most recent advances in game theory, we summarize here the most salient characteristics of the Classical Game Theory.

2.3 *Principles of Classical Game Theory*

We recall first the three most common models of classical game theory, widely described and studied in most books on game theory:

1. The Prisoner's Dilemma (PD): This is a paradigm of noncooperative game, which mathematically models how players' mutual cooperation with a better collective outcome/payoff is not a Nash Equilibrium of the game. But instead, mutual betrayal/defection leads to the unique Nash Equilibrium, which is not event Pareto-optimal. That is, the tension between individual rationality and collective benefits. Its applications include Price Wars and Trade Agreements, Cartels operations, Arms races, Plea Bargains, Machine Learning and Artificial Intelligence Ethics. The PD has the following general payoff matrix

		Column Prisoner	
		C	D
Row Prisoner	C	(R,R)	(S,T)
	D	(T,S)	(P,P)

with $T > R > P > S$, and classically, T is the temptation to defect, R the reward for mutual cooperation, P the punishment for defection, and S the so-called "sucker"'s payoff to the player who defects while the other cooperates.

2. Matching Pennies: This models zero-sum games with no pure Nash Equilibrium, with the typicall payoff

		Column Player	
		H	T
Row Player	H	$(1,-1)$	$(-1,1)$
	T	$(-1,1)$	$(-1,-1)$

The optimal strategy is mixed, i.e., randomization with equal probability, leading a mixed Nash Equilibrium. Its applications include Markets Competition, Bidding Wars, Hacker Vs. Defender, Submarine warfare, and Soccer Penalty Kicks.

3. The Battle of the Sexes: A paradigm for coordination games, with a payoff matrix such as

		Husband	
		O	F
Wife	O	$(3,2)$	$(0,0)$
	F	$(0,0)$	$(2,3)$

Players with conflicting individual preferences benefits from being together. It has two pure Nash Equilibria and one mixed Nash Equilibrium. It can be applied to Communication and Negotiation scenarios, Trade Agreements Military Alliances, Business and Technology Standards,

Remark 2.15 As presented in the subsequent sections, these games will greatly benefit from straight quantumization and p-adic quantumization as described below.

Classical game theory, in particular, in its multi-player strategic form, mathematically describes the strategic interaction, the *game*, conflictual or cooperative, between two or more entities, the *players*. A player's *pure strategies* refer to their ability to interact with the other players, and their stakes in the game are their *payoffs or utilities,* which, as *rational entity*, they seek to optimize/maximize, consistent with some preferences ordering over their payoffs.

A *game play* calls for the choice of a pure strategy by each player, leading to a tuple of pure strategies, the *pure strategy profile*, which determines the player's payoff, under the assumption that each player's choice of pure strategy is indeed their best response to the others' choice. The outcome of choosing and using such pure best response strategy is the pure strategy *Nash Equilibrium* (PNE), a fundamental goal of multi-player game theory.

Importantly, the appearance, if any, (as it may not even exist) of a PNE does not imply *uniqueness or optimality*. Players may (and in general, they must) therefore extend their strategy set to include *mixed strategies,* (von Neumann) a randomization between their pure strategies through an efficient probability distribution over the pure strategy set, and yielding *expected payoffs*.

A fundamental result for the strategic form (with a finite strategy set) multi-player game is Nash's famous theorem, we recall as

Theorem 2.16 (Nash Existence Theorem) *If each player in a multiplayer game (two or more players) has a finite number of pure strategies, then the game has a Nash Equilibrium (not necessarily unique) in (possibly) mixed strategies.*

The extension of a game strategy or payoff profile can take several possible directions. In the classical case, the co-domain of the payoff function is the real numbers field, $\mathbb{R}$, which can be switched to the p-adic numbers field $\mathbb{Q}_p$ described below. Another possible extension, exposed below, is the use of quantum strategies, initiated by Meyer in [55].

Additional extensions are also presented below in the sections on Intelligent Games and Non-Archimedean or p-adic Games.

3 Intelligent Game Theory

In addition to the above categories of games, the so-called Intelligent Game Theory (IGT) is a fairly recent extension. It combines game theory with the advancement of machine learning (ML), artificial intelligence (AI) and computational methods to improve decision-making in strategic environments, to learn, adapt and optimize accordingly. This leads to applications in robotics, autonomous systems, multi-agents systems and cybersecurity and many other emerging areas.

3.1 Multi-agents Systems and Learning in Game Theory

Constructing agents that exhibit intelligent behavior is an expanding endeavor in research on computer systems, artificial intelligence, agent-based software development. So intelligent agents refer to smart programs able to learn from their environments and user interactions to perform tasks and make their own decisions in order to maximize overall benefits (utility-based agents), to achieve a common goal (multi agents in transportation systems, robotics, social networks, …).

Intelligent Game Theory aims at designing efficient strategies for such agents to leverage their own decisions against the actions of other agents in a dynamic and potentially adversarial environment.

Through Reinforcement Learning, these agents learn optimal strategies using trials and errors, rewards and penalty in order to continuously adapt their strategies and optimize outcomes. This is especially useful in competitive games, traffic management, network optimization.

To handle high-dimensional state spaces and higher order complexity the intelligent agents approximate complex game-theoretic models using advanced ML techniques such as deep neural networks.

It goes without saying that intelligent agents operate under conditions of uncertainty or incomplete information. Bayesian game theory, approximate inference methods, belief-based learning assist these agents to optimize tier strategies in such environments. AI-driven strategies can help predict and optimize outcomes including in iterated, repeated and incomplete games.

This is to say that there are ample and varied opportunities to apply Intelligent Game Theory. We name here the following:

- Social Networks: we now live in the age of expanding social networks on platforms such as YouTube, Instagram, Facebook, X/Twitter, Tik-Tok. AI assisted Game Theory is used to model interactions such as the so-called *reputation systems* where the actions and behaviors of others influence and impact sometimes negatively most users' behavior. Combining game-theoretic models with AI assisted recommendation systems and sentiment analysis will help predict and influence user behavior.
- Healthcare: This is one important area of application of Intelligent Game Theory. Decision-making systems involve a variety of stakeholders: patients, doctors and their medical assistants, medical insurers. Here intelligent agents learn optimal strategies for resource allocation in medical treatments. This will lead to a complete and positive overhaul of the healthcare system, in particular, in countries with no universal efficient healthcare system.
- Economic markets: Game theory is being assisted by machine learning (ML) and AI to analyze auctions, pricing strategies, market competition for fairness and efficiency, prediction of market movements and optimization of resource distribution in supply chains.
- Cybersecurity: This is certainly a timely area of application. IGT efficiently models interactions between attackers and defenders, with optimal defense strategies against potential cyber attacks to predict and mitigate threats in intrusion detection systems. Of particular importance is the design of effective protocols and decision-making mechanisms in adversarial security settings, including airport and seaport patrols on the models of algorithmic game-theoretic models for K9-patrols at LAX and national parks protection against poachers.
- Autonomy: Game-theoretic models are definitely useful for autonomous vehicles to avoid collision, optimize traffic flow, negotiate with other vehicles. Task allocation and path planning problems for multi-robot systems can be solved through intelligent cooperative game theory. Overall, coordination and decision-making of autonomous agents in multi-agent environments greatly benefit from the development of intelligent game theory.

3.2 *Testing the Efficiency of Intelligent Game Theory*

- Computing equilibrium solution and in particular Nash Equilibrium involves some intractable computational complexity. Approximate Nash Equilibrium can be computed using some learning algorithms such as Q-learning or deep Q-networks with intelligent agents.

- in Intelligent Cooperative Game Theory, AI is an efficient tool to compute the well-known solution concept of Shapley Value which allocates payout based on contributions, in particular, in distributed systems or collaborative robotics with intelligent agents working together to complete a target task while deciding on fair compensation for each contribution.
- Computational complexity is inherent to very large scale problems, which necessitates the use of efficient computation methods such as deep learning and neural networks to find optimal strategies for game-theoretic solution in multi-agents environments.
- Scalability presents another challenge as the number of intelligent agents increases, along with an exponential growth of the complexity of intelligent game-theoretic models.
- Another critical challenge is of course coordination and cooperation of intelligent with any central authority to achieve specified goals.
- We should note here also the importance of ethics considerations in all these applications from market design, to autonomous systems to healthcare systems all requiring fairness and transparency.

4 Differential Game Theory

Classical game theory has been extended to the dynamic systems modeled by differential equations with strategies evolving over time. Differential Game Theory (DGT)[18] deals with the study of dynamic interactions between players in continuous time, and models situations where players have a certain level of control over the rate of change (differential) of certain variables over time, such as the speed of a vehicle or the flow of resources in a network. Therefore such a game theoretical approach is often used in engineering, economics, and ecology to analyze optimal control problems, resource management, and environmental policy. That is, whenever there is a dynamical optimization or optimal control with strategic interaction. For example

- Aerospace Engineering: The design of optimal guidance and control systems for spacecraft and aircraft will draw from differential game theory taking into account factors such as fuel consumption, aerodynamic performance and environmental constraints. For example, an aircraft landing subject to wind disturbance. Air traffic control.
- Control Systems and Robotics including autonomous vehicle coordination.
- Ecology: the study of the dynamics of populations and ecosystems, assuming individuals have control over their reproductive rates or foraging behaviors. The

[18] The introduction of Dynamics Games is due to Rufus Isaacs, 1951, followed with the 1965 book *Differential Games: A Mathematical theory with applications to warfare and pursuit, control and optimization,* Other later contributors include A. Merz and Breakwell, T. Basar et al.

differential game theoretic approach would contribute to identity optimal management strategies for conservation, pest control, or fisheries.
- Military and Defense where there is a need to model competitive strategies between adversaries in conflict situations.
- Economics and Management Science: Differential Game Theory is effective in analyzing strategic interactions between firms in dynamic market, where production rates, pricing strategies, and investments in research and development are traditionally controlled by the firms.

Differential game theory can be seen indeed as an extension of optimal control where game theory provides equilibrium concepts including Nash equilibrium and Stackelberg equilibrium. See Dockner et al. [26]. Solution techniques in optimal control include the *Pontryagin maximum principle* and *dynamic programming*. See Bellman's principle of optimality, Bellman et al. [10]. The theory of differential game adds the concept of information structure: *open loop* to mean strategies are function of time only, and correspond to the maximum principle, *closed loop no memory* shows up in dynamic programming for strategies function of time and the current state, and *closed loop memory* for strategies function of time, current and past states.

The types of differential games are determined by the information patterns $\eta_i(t)$ available to player P_i at time t and the payoff structures: open-loop information or closed loop information. To illustrate, in a n-players differential game with infinite horizon and state x each player P_i has a control u_i and optimizes an objective function subject to a differential equation. For instance

$$\max_{y(.)} \int_0^{\infty} e^{-rt}[\beta\gamma(t) - 0.5y^2(t) - 0.5\gamma s^2(t)]dt, \tag{4.1}$$

subject to

$$\dot{s(t)} = y(t) - \delta s(t), \quad s(0) = s_0$$

with r the discount rate and δ the natural assimilation rate.

5 Algorithmic Game Theory

A rapidly developing sub-game theory, *Algorithmic Game Theory* (AGT) refers to games involving the use of algorithms or any computational protocol or mechanism for decision-making. That is, the study of game-theoretic problems for the "algorithm" and "computational" perspectives: computing equilibria in an efficient, fast, possibly distributed and centralized manner; thus the question: what is the complexity of computing a Nash Equilibrium? In other words, how long does it take until selfish, however rational, players converge to an equilibrium? Daskalakis, Goldberg, and Papadimitrion, Chen and Deng showed in 2006 that convergence to equilibrium can be prohibitively long. In particular under Nash's fundamental theorem stating that *Every finite non-cooperative strategic game of two or more players has a (mixed) Nash equilibrium.*

Algorithmic methods have been used extensively for potential and congestion games to compute their pure Nash equilibria. However, in most experiments the running time has been "exponential". See in reference Nisan et al. [62].

The key in AGT is on understanding how computational algorithms interact with game-theoretic frameworks:

- Computing Nash equilibria in large scale systems. This can be computational difficult, leading to research on approximation algorithms and efficient mechanisms.
- Driving bidding strategies and optimization in Auction systems[19] (e.g. eBay or Google Ads auctions).
- Managing traffic routing by selecting the best routes or service allocation strategies.
- Determining the assignments of resources on platforms such as online marketplaces, ride-sharing apps matching drivers with riders in order to reach a computational equilibrium.

We now present one of the most prominent applications of Algorithmic Game Theory, *Security Games*.

5.1 Security Games

Game-theoretic methods are applied to the modeling and analysis of security scenarios, involving mostly attacker and defenders interacting strategically with the goal to optimize security measures while accounting for adversarial behaviors. The scenarios include: cybersecurity, military defense, counter-terrorism, patrolling seaports and airports, protection of critical infrastructures and resources, protecting wildlife and forest form poacher and smugglers, curtailing the illegal flow of weapons, drugs and money across international borders.

The science of security games amounts to designing game-theoretic models to predict and counteract the actions of attackers and optimize the use of available resources for defense. Here a Nash equilibrium is reached whenever neither the attacker nor the defender can improve their payoff by changing their strategy/plan given the strategy/plan put in place by the other, under the rationality assumption.

Mixed strategies are of great importance in this class of games, in particular, for large and complex systems. Actions or plans are chosen or design with certain probabilities, e.g., random patrolling of different routes in order to make it harder for the attacker to predict where security is strongest. Indeed, one of the central questions is how to efficiently allocate finite resources to maximize protection against potential

---

[19] A fundamental process of buying and selling goods and services, an auction transfers/exchanges goods and services through taking bids as follows: seller wants to sell an item; bidder P_i has a valuation v_i utility of P_i is v_i-price paid, and 0 if loses auction submit bid to maximize utility. Seller selects the higher bidder.

attackers (e.g. security personnel, budget, sensors), in order to balance the cost of security measures with the potential losses from a successful attack.

We distinguish two main classes:

- Stackelberg security games[20] with asymmetric relationship between player and most common in the real-world: Defender commits first to strategy, a course of action before any attack (e.g., allocate security resources) and then the attacker reacts to the defender's strategy, by endeavoring to exploit the vulnerabilities in the defensive strategy.
  For example the Stackelberg model has led to the development of new algorithms providing randomized patrolling and inspecting strategies. Examples include ARMOR deployed at LAX in 2007 to randomized checkpoints; IRIS for randomized deployment of US Federal Air Marshals since 2009; PROTECT for randomized patrolling of ports by US Coast Guard (port of Boston since 2011 and port of New York since 2012, spreading to other ports in the nation). PAWS tested by rangers in Uganda for protesting wildlife in national parks. MIDAS for the US Coast Guard for the protection of fisheries.
- Bayesian security games in which players have totally incomplete information about the other players' strategies or payoffs. Mixed strategies are therefore determinant as defender and attacker consider the probabilities of various types of attack and make decisions based on the belief about the adversary's actions.

This is to say that the science of security seeks to effectively combine game theory, in particular algorithmic game theory, optimization, strategic thinking oftentimes through "war gaming" and regular drills, to model and solve real-world security issues. This field of research endeavors is crucial in modern days for cybersecurity, military defense, infrastructure protection, counter-terrorism, and strives to leverage the advent of artificial intelligence, machine learning and quantum computing. The most relevant challenges include: complexity, uncertainty, oftentimes due to incomplete information. Dynamics nature of threats requiring constant updates of strategies over time. Ethical implications: for instance, surveillance overreach compromising individuals' privacy, counter-terrorism's impact of civil liberties.

6 Quantum Game Theory

Recall that "mixed games" resulted from extending the domain of "pure" payoff profile to include probability distributions. The idea of such extension can be pursued in other directions, including in the non-Archimedean/p-adic and the quantum directions. Game Theory is being extended to include the so-called *Quantum games* in which players have access to quantum systems or quantum effects while deciding on their strategies. The most important quantum effects include superposition and entanglement and/or the randomness of quantum measurements. Strategies subjected

[20] First introduce to model leadership and commitment by Kiekintveld et al.

to these effects cannot be emulated in classical games. It should be mentioned here some of the pioneering works: first, the 1999 seminal paper by Myers showed that a "quantum player" always wins against a classical player. The Pareto-optimality of a Nash equilibrium compared to the classical Nash equilibrium was showed by Eisert, Wilkens and Lewenstein in 2000 by quantumizing a two-person game with both players having access to quantum effects. And according to Dahl and Landsburg (2011) a quantum payoff resulting from using quantum strategies is at least as great as the payoff using classical strategies.

6.1 *Quantum Strategy Profile*

Meyer first extended players' pure strategies to include *pure quantum strategies*, i.e., the "quantumization" of the game, with the appearance of "new" (almost) optimal Nash Equilibria in terms of quantum strategy profiles. A further extension may be considered to include *mixed quantum strategies*, i.e., a probability distribution over the pure quantum strategies.

A Quantum Primer

Superposition and entanglement are the most distinguishing features on quantum systems: Superposition refers to a system' ability to appear in multiples states simultaneously whereas entanglement describes a strong (non-classical) correlation between quantum systems with no classical analogue.

We first present some of the key principles of quantum information

1. Principle 1: A quantum physical system corresponds a complex Hilbert space $\mathbb{H}$, that is, a complete Inner Product space $\langle \mathbb{H}, <.,.> \rangle$ over the complex number $\mathbb{C}$.
2. Principle 2: The state $|\psi\rangle$ of a quantum system is given by a unit vector $v \in \mathbb{H}$ of the associated Hilbert space.
3. Principle 3: For every orthonormal basis $\mathscr{B} = (b_i)_{i=1,\dots,n}$ in the Hilbert space, there is an associated measurement[21]: the outcomes of the measurement are in $\mathscr{B}$. The probability of outcome $x = \sum_{i=1}^{n} < b_i, x > b_i \in \mathscr{B}$ when the system is state $v \in \mathbb{H}$ is $\| < x, v > \|^2$ (*Born rule*). If the outcome of the measurement is x then after measurement the system will be in state $\frac{<x,v>}{\|<x,v>\|}x$ (*Wave function collapse*).
4. Principle 4: Time Evolution $E : S(\mathbb{H}) \rightarrow S(\mathbb{H})$ is linear, with $S(\mathbb{H})$ the set of states in $\mathbb{H}$. (E is the operator). The linearity implies the preservation of superposition, e.g., $E(a|0\rangle + b|1\rangle) = aE(|0\rangle) + bE(|1\rangle)$.
5. Principle 5: There is a measurement associated with every complete family of orthogonal projections.[22]

[21] A measurement can be thought of as an abstract physical process implemented to always return an outcome $\mathscr{O}$, depending on the state in which the system is. It is probabilistic.

[22] A Projection P is a linear operator with $P^2 = P$. A set of projections $\{P_1, \dots, P_n\}$ such that $P_i P_j = 0$ and $\sum P_i = \mathbf{1}$ is called a complete family of orthogonal projections.

6. Principle 6: For $\mathbb{H}_1$ and $\mathbb{H}_2$ two Hilbert spaces representing tow quantum physical systems, the Hilbert space of the join system is the tensor product $\mathbb{H}_1 \otimes \mathbb{H}_2$. The above family is a projective measurement if the projections are also self-adjoint $P^* = P$.

The quantum state of an isolated system is described by a vector in a complex Hilbert space $\mathbb{H}(C)$, usually finite dimensional. qubits are 2-dimensional systems, and the quantum analog of the classical "bits", represented, using Dirac's famous braket notation

$$|\psi\rangle = \begin{bmatrix} \alpha \\ \beta \end{bmatrix}, \quad \alpha, \beta \in \mathbf{C}.$$

The particular states of qubits are

$$|1\rangle = \begin{bmatrix} 0 \\ 1 \end{bmatrix} \quad |0\rangle = \begin{bmatrix} 1 \\ 0 \end{bmatrix}$$

leading to the general quantum of a qubit as a complex linear combination of these canonical states

$$|\psi\rangle = \alpha\,|0\rangle + \beta\,|1\rangle\,,$$

that is, a superposition state, a defining characteristic of a quantum system with non classical analogue.

We label $\{|0\rangle\,, |1\rangle\}$ a set of two normalized, mutually orthogonal states, for qubit, defining these states:

$$|1\rangle = \begin{pmatrix} 0 \\ 1 \end{pmatrix} \quad |0\rangle = \begin{pmatrix} 1 \\ 0 \end{pmatrix}$$

The state of the qubit is represented either in a classical or a superposition state

$$|\,|\Psi_{classical}\rangle = \gamma\,|n\rangle\,, \quad n = 0, 1 \quad |\gamma|^2 = 1 \tag{6.1}$$

$$|\,|\Psi_{superposition}\rangle = \alpha\,|0\rangle + \beta\,|1\rangle \quad |\alpha|^2 + |\beta|^2 = 1 \tag{6.2}$$

Note that when observed the superposition collapses in the state in which it is observed.

A system state can be also represented by positive, semi-definite, Hermitian matrices of trace 1 called density matrices.

The probability distribution is *separable* when each player can choose their mixed strategy (i.e., through probabilities) independently of each other. However, there are game scenarios in which the probability distribution is not separable, with strategies *correlated*. These are the so-called correlated game.

Quantum Game Play: Quantumizing The Battle of the Sexes

Below we illustrate using one of well-know classical games, the Battle of the Sexes.[23]

The game play: A *married couple*, Husband (H) and Wife (W) are planning a weekend outing "together" (keyword). Madame prefers to go to the Opera (O) while Monsieur (H) would prefer to go a football game (F). We also assume the following utility increment: 1 if they go to the event of their preference and 0 otherwise. Each gets an increment of 2 for going to the event together and 0 for going separately. Therefore the game play has: a set of 2 players $\{W, H\}$; a set of 2 *pure strategies* $\{F, O\}$ and the following payoff/utility matrix

		Husband	
		O	F
Wife	O	(3,2)	(0,0)
	F	(0,0)	(2,3)

We then proceed with solving the game.

1. This game has clearly two pure Nash Equilibria at (F, F) and (O, O). This prompts the need to randomize.
2. We look for a mixed Nash Equilibrium: The Wife chooses Football with probability p and the husband with probability q., leading to the updated mixed matrix. Upon randomization by both Wife and Husband to reach indifference between going to Football and Opera (i.e., equal payoff for going to either event), we then solve to obtain $p = \frac{2}{5}$ and $q = \frac{3}{5}$. As these are independent probabilities, the probability that both go to the football game is the product, i.e., $\frac{6}{25}$.
   Similarly, we have both at the opera with the same probability $\frac{6}{25}$, but husband at the football while wife at the opera has the probability of $\frac{9}{25}$ and husband at the opera while wife at the football has the probability of $\frac{4}{25}$.
   Therefore the mixed Nash Equilibrium $\sigma^* = (\sigma_1^*, \sigma_2^*)$ is given by
   $$\left(\frac{2}{5}, \frac{3}{5}\right); \quad \left(\frac{3}{5}, \frac{2}{5}\right).$$
   Consequently this game exhibits three Nash Equilibria, consisting of 2 pure and 1 mixed, prompting the concern of how the players should decide. This indicates that simple solutions are not always available even in the simplest game play. The most likely outcome is (F, O) if both decide to randomize with a probability of $\frac{9}{25}$. Coordination would be a better outcome for the married couple.
3. This game play may be extended to an evolutionary game scenario.
4. We now consider the extension to this game to a quantumization of the play.

In the quantum version the players Husband and Wife use quantum strategies that are superposition and entanglement of classical strategies The game is quantumized

[23] The game first appeared in the book "Games and Decisions" by Duncan Luce and Howard Raiffa in [53].

by choosing quantum qubits systems given by 2 dimensional Hilbert spaces $\mathbb{H}_w$ and $\mathbb{H}_h$ associated with the players wife and husband.

Then we consider their *tensor product* $\mathbb{H}_w \otimes \mathbb{H}_h$ with it orthonormal basis denoted using the classical pure strategies O and F, now considered as vector states $|o\rangle$ and $|F\rangle$

$$B = (|OO\rangle, |OF\rangle, |OO\rangle, |FO\rangle, |FF\rangle$$

We proceed using the quantum formalism recalled above in the primer. In particular, we deal with *density matrices* associated with the strategies. For example, wife and husband have now at their disposal the following *entangled state*

$$|\psi\rangle = \alpha |OO\rangle + \beta |F\rangle, \quad |\alpha|^2 + |\beta|^2 = 1$$

with the associated density matrix ρ

$$\rho = |\alpha|^2 |OO\rangle |OO\rangle + \alpha\bar{\beta} |OO\rangle |FF\rangle + \bar{\alpha}\beta |FF\rangle |OO\rangle + |\beta|^2 |FF\rangle |FF\rangle$$

In this particular game of the Battle of the Sexes, one exploits the entangled strategies to obtain a Nash Equilibrium Pareto-optimal as the unique viable solution of the game (giving an even higher reward to both wife and husband than any other solution): in summary, a quantum strategy exhibit a more attractive solution than the classical strategy. However, it should be noted that only using entangled strategies allow the players wife and husband to reach a unique solution, both getting the same expected payoff. The interpretation is that entanglement of the strategies O and F actually represents a strong correlation, directing wife and husband to act the same way. That is, wife and husband, by entangling their strategies, must now play the same strategy with the option to choose sometimes opera (O) and sometimes football (F), obtaining the best reward playing O one-half of the times and playing one-half F. This is indeed more realistic for a married couple, as if marriage, a strong correlation, is here "quantumized into entanglement"! See more details on the computation in references Marinatto and Weber [54], Sowmitra [22].

Remark 6.1 (*Commnents*) There is a considerable literature on quantum game. Classical game theory has been applied for various modern decision-making processes from diplomacy to economics to cultural dynamics to national security and to agricultural and biological sciences. For example, the game of chicken and the cuban missile crisis. As a recent extension of the classical game theory quantum game theory is seeing many practical applications, leveraging the superposition and entanglement features available in quantum games. One can also mention work on quantum routing game.

7 Non-Archimedean Game Theory

Non-Archimedean refers to the violation of the Archimedean Principle[24]. It suffices to verify that there are no infinitesimals elements to confirm the Archimedean Principle. In other words, a valued field $< \mathbb{K}, |.| >$ is Archimedean if

$$\forall x \in \mathbf{K}^*, \exists \quad n \in \mathbf{N} \quad | \quad |\underbrace{x + \ldots + x}_{n-times}| = |nx| > 1.$$

Classical game theory is studied in an Archimedean/Euclidean space. Therefore Non-Archimedean game theory is studied using non-archimedean spaces, in particular, through p-adic analysis.[25] Archimedean absolute value $|.|$ satisfies the *Triangle Inequality*, whereas the p-adic absolute value satisfies the ultrametric (strong) inequality

$$|x + y| \leq \max(|x|, |y|).$$

Such an inequality leads to a peculiar geometric space where among other things, all triangles are isosceles, every point in a disc is also a center of the disc, all open balls are also closed (clopen balls). Here also there are non-constant functions with zero derivatives (called pseudo-constants). The p-adic space is totally disconnected. The non-archimedean or p-adic approach handles beautifully hierarchical and fractal-like structures. It goes without saying that applying p-adic analysis to game theory requires a strong understanding of p-adic number system and all game theory concepts.

7.1 p-Adic Primer

Here is a p-adic refresher sufficient for this compendium. The set of real numbers $\mathbf{R} = \mathbb{R}_\infty$ is indeed a completion of the rational numbers $\mathbb{Q}$ with respect to the Euclidean or Archimedean norm denoted here $|.|_\infty$ which satisfies the well-known Triangle Inequality $|x + y|_\infty \leq |x|_\infty + |y|_\infty$

From the canonical representation of a positive integer $n = \prod_{i=1}^{s} p_i^{k_i}$ and the normalized form of a non-zero rational $r = p^k \frac{a}{b}$, p prime with a and b co-prime, the p-adic absolute $|.|_p$ is defined as

---

[24] Given by Otto Stolz in the 1880s in honor of the ancient Greek Geometer and Physicist Archimedes of Syracuse who stated the principle as Axiom V in his *On the Sphere and Cylinder*.

[25] p-adic numbers, ultrametric spaces, non-archimedean numbers, isosceles spaces all express the same idea. Kurt Hensel initiated the p-adic analysis in 1897.

$$|r|_p = \begin{cases} p^{-k} \text{ for } r \neq 0 \\ 0 \text{ for } r = 0 \end{cases} \tag{7.1}$$

The p-adic absolute value $|.|_p$ or ultra-norm and its induced p-adic distance $d_p(x, y) = |x - y|_p$ satisfy the *Ultrametric or Strong Inequality*

$$|x + y|_p \leq \max(|x|_p, |y|_p) \tag{7.2}$$
$$d_p(x, z) \leq \max d_p(x, y) + d_p(y, z) \tag{7.3}$$

The inequality changes to equality for $|x|_p ne|y|_p$ or $d_p(x, y) \neq d_p(y, z)$ In other words, all triangles in the p-adic space are isosceles.

The completion of $\mathbb{Q}$ with respect to the p-adic norm $|.|_p$ gives the set of p-adic numbers $\mathbb{Q}_p$ with its associated series expansion form

$$\mathbf{Q}_p \ni r = \sum_{k=\nu}^{\infty} r_k p^k | \quad r_k \in \{0, 1, \dots p - 1\} \subset \mathbf{Z}. \tag{7.4}$$

The expansion is also symbolically represented using the "decimal" or *radix* point (as in the real case) by $r = r_\nu \cdots r_{-1} \bullet x_0 x_1 \cdots x_n \cdots$.

The set $\{0, 1, \dots p - 1\}$ is the set D_p of digits in the representation of p-adic numbers. The set $\mathbf{Q}_p$ is the field of fractions of $\mathbf{Z}_p$ the space of *p-adic integers*, the compact unit disk $\mathbf{Z}_p = \{x \in \mathbf{Q}_p| \quad |x|_p \leq 1\}$.

Other peculiarities on the p-adic analysis include

- The set $\mathbb{Q}_p$ is locally compact and totally disconnected: That is, calculus is performed with no expectation of "reasonable" analyticity.
- While a natural geometric ordering for the usual Euclidean metric is given by the real number line, in the ultrametric case a hierarchical tree provides such an ordering.
- Given a p-adic infinite expansion $\mathbb{Z}_p \ni x = x_0 + \cdots + x_{n-1} p^{n-1} + O(p^n)$, we have $\bar{x} = x_0 + \cdots + x_{n-1} p^{n-1}$ as its p-adic approximation, i.e., $x \equiv \bar{x} \quad mod \quad p^n$. And $|x - \bar{x}|_p \leq p^n$ ensuring convergence. n is the *order* or *absolute precision* of $\bar{x}$, with the *relative* precision given by $n = \min\{i \in \mathbf{Z}, n_i \neq 0\}$.
- Relevant to game theory, in particular, the computation of Nash Equilibria, is the fact that **p-adic errors do not add**:
  That is, $x + O(p^{k_1}) + (y + O(p^{k_1}) = x + y + O(p^{\min(k_1, k_2)})$. This is a tremendous advantage over the real precision with its compounding of round-off errors [24, 36, 51, 69].
- For $\nu_p(x) := \max\{k \in \mathbf{Z} : p^k | x\}$ called the p-adic valuation of x, we also have $x + O(p^{k_1}) \times (y + O(p^{k_1}) = x \times y + O(p^{\min(k_1, \nu_p(y) k_2, k_2 + \nu_p(x))})$.
- p-adic computation also involves the so-called Hensel Code originated from the Hensel's Lifting Lemma [35, 45, 65].

7.2 Applying p-Adic Analysis to Game Theory

In light of all the above applying p-adic analysis to game theory is indeed justified.

- For one, preferences in decision-making are modeled by utility/payoff functions, which are traditionally real-valued, but can, alternatively, be expressed using p-adic numbers, to account for a different scale of the idea of "closeness" or "distance" between choices or strategies, in particular, in situations where relative versus absolute difference between strategies proves more relevant.
  That is, to leverage the ultrameticity of p-adic topology, i.e., using the ultrametric distance between strategies instead of the Triangle Inequality distance. In some instances, p-adic valuations may provide a better working framework for decision-making processes with imperfect information.
- As recalled, p-adic or ultrametric spaces have peculiar geometry and topology, which could be more suitable to analyze the strategy space in games; indeed, the ways strategies are related to each other certainly influences the equilibrium or any outcome of the game. Note that p-adics are closely tied to local structures, thereby emphasizing local differences in payoffs.
- Evolutionary Game Theory (EGT) is classically studied in Euclidean spaces. One may now consider a p-adic evolutionary game theory: mutations, which play important role in EGT can now be thought as discrete steps appearing according to some p-adic valuations, with impactful changes and dramatic evolutionary shifts resulting from very small mutations in the close vicinity of the strategies.
- In EGT, there is a potential of players refining their strategies to improve their relative payoffs. A p-adic evolution could model such refinement process, to include a player's strategy that evolves in a p-adic neighborhood with gradual adjustment to optimal and desirable strategies.
- Notably, an evolutionarily stable strategy (ESS) can be evaluated through an ultrametric distance instead of the classical Euclidean/Archimedean metric, recalling that p-adic closeness differs for the Euclidean one. For example, 2 is 7-adically closer to 51 than it is to 1 because $d_7(2, 51) = i/19$ whereas $d_7(1, 2) = 1$.
  Or 3-adically one can express -1 as $-1 = 2 + 2.3 + 2.3^2 + \cdots + 2.3^n + \cdots$.
  Therefore p-analysis may lead to new evolutionary equilibria with a Nash Equilibrium.
- Recall the hierarchical structures of p-adic spaces. This could be indeed used to describe different layers of strategy evolution, contrasting localized evolutionary changes within smaller subgroups in large populations with the broader evolutionary dynamics of the entire population. This will better describe the real-life interplay between local and global evolutionary pressures, For examples, in the socio-cultural-economic scenario below, the interplay between local Nash Equilibria and the central Nash Equilibrium: Small changes in local subpopulations of a highly stratified society, such as the American society, may easily propagate across the entire society, possibly directing or redirecting overall strategy shifts in a highly nonlinear manner (not excluding chaotic one).

- p-Adic Probability Distribution: In classical game theory and its related EGT, mixed strategies are considered using a classical probability distribution of the state vectors. p-adic analysis leads to a development of a theory of p-adic probability taking values in the p-adic numbers.: frequencies, relative or otherwise, take place in the set $[0, 1] \subset \mathbb{Q}$ whose closure is the same set in the Archimedean topology but whose closure in the p-adic topology is the whole set $\mathbb{Q}_p$ of p-adic numbers. Consequently, all possible sequences in $\mathbb{Q}_p$ summing up to 1 are considered legitimate p-adic probability distributions. For example, a sequence such as $\{1, -5, -1, 6\}$ which is certainly not a standard probability distribution. That is, in the p-adic probability values greater than 1 even negative are allowable.

8 p-Adic Quantum Game

The p-adic analysis has received many concrete applications in recent years, in particular, applications to physical theories; due to the existence of the *Planck length* $l_p \approx 10^{-35}$ in quantum gravity, many non-Riemannian model were being considered, to include the models based on the non-Archimedean field $\mathbb{Q}_p$ of p-adic numbers, which exhibits a fractal-like, ultrametric and hierarchical structure, as recalled above. A particular important application has been in computer science and cryptography pertaining to the generation of the so-called *pseudorandom numbers* and *uniform distribution of sequences.*

There is also a great interest in developing a p-adic quantum information theory alongside the quantum information theory. However, we present here some preliminary ideas and work on **p-Adic Quantum Game Theory**, a theory we are actively developing drawing from all the above combining the quantum and p-aidc primers, with the goal to leverage the distinguishing features of the p-adic analysis and the quantum approach.

To recap, quantum game features *superposition and entanglement* of players' strategies, leading oftentimes to a "better" and more efficient Nash equilibrium as in the case of the quantum Battle of the Sexes.

On the other hand, there is a tremendous advantage in considering a p-adic analysis of games, in which players' payoff function has now a co-domain in the p-adic number system $\mathbb{Q}_p$. A p-adic neighborhood of players' strategy accommodates best some gradual adjustment to optimal strategies in evolutionary game theory, as well as the consideration of different layers of strategy time-evolution. Let also mention that one could successfully leverage the p-adic probability and a p-adic time in evolutionary game theory. Note that in classical evolution theory, time is Euclidean/Archimedean, i.e. classical time is merely statistical (topologically and metrically like the real line), whereas a p-adic time could better describe the circularity of time periods (the non-linearity of time viewed as a series of repeating cycles).

8.1 Tenets of p-Adic Quantum Game

The formulation of standard quantum theory relies fundamentally on the field $\mathbb{C}$ of complex numbers, which can also be regarding as a quadratic extension of the real field $\mathbb{R}$. The complex conjugation is the natural involution on the complex numbers field. The fundamental requirement of the p-adic quantum game, is to replace the field $\mathbb{C}$ with a quadratic extension of $\mathbb{Q}_p$. We follow here the approach taken by Aniello et al. See Aniello [3] for more details. That is, we consider a non-quadratic element in $\mathbb{Q}_p$, i.e., $\mu \in \mathbf{Q}_p \quad | \quad \mu \notin (\mathbf{Q}_p)^2$ to finally define

$$\mathbf{Q}_p(\mu) := \{z = x + y\sqrt{\mu} \quad | \quad x, y \in \mathbf{Q}_p\} \tag{8.1}$$

$\mathbf{Q}_p(\mu)$ is a field extension of $\mathbf{Q}_p$, a 2-dimensional $\mathbf{Q}_p$ vector space, endowed with a non-Archimedean valuation

$$|z|_{p,\mu} := \sqrt{|z\bar{z}|_p} \tag{8.2}$$

with $z = z = x + y\sqrt{\mu}, \quad \bar{z} = x - y\sqrt{\mu}, \quad z\bar{z} = x^2 - \mu y^2 \in \mathbf{Q}_p$.

There are many non-isomorphic quadratic extensions of $\mathbf{Q}_p$; for example, for $\eta = -1$ and prime $p \equiv 3(mod\, 4)$ is a non-quadratic element of $\mathbf{Q}_p$ (e.g., for $p = 7$, take $\mu \in \{-1, 7, -7\}$. For $p = 2$ take any $\mu \in \{2, 3, 5, 7, 10, 14\}$.

Once a quadratic extension has been determined, we then define a p-adic Hilbert space $\langle \mathbb{H}_p, \|.\|, < ., . > \rangle$ where, $< ., . >$ is a *non-Archimedean inner product* and $\|.\|$ is an *ultranorm* such that, in general, the ultranorm does not stem directly from the inner product.

Armed with such a definition of p-adic Hilbert space, for example, a p-adic qubit is a quantum system described by a 2-dimensional p-adic Hilbert space.

Remark 8.1

- In addition to the fact that for p-adic Hilbert space, the ultranorm is not directly coming from the inner product, i.e., in general we have $\|z\| \neq \sqrt{| < z, z > |}$, the space may contain non-vectors z with $< z, z >= 0$, i.e., z is an *isotropic* vector. For example, $z = b_1 + \sqrt{-1}b_2$, where b_1, b_2 are in an orthonormal basis of $\mathbb{H}$.
- Here p-adic quantum states are p-adic statistical operators defined as linear operator $\rho \in \mathscr{L}(\mathbb{H})$ such that $\rho = \rho^*$ and $trace(\rho) = tr(\rho) = 1$, i.e., a trace one self-adjoint linear operator forming the set $\mathscr{S}(\mathbb{H})$.
- The description of p-adic quantum measurement (i.e., the *observables*) is given by self-adjoint operator valued measures (SOVM), a family $\mathscr{M} = (M_i)_{i \in I}$ (I a finite index set) of self-adjoint operators in $\mathbf{H}$ such that $\sum_I M_i = \mathbf{I}$.
- We also define a linear functional $\omega \in \mathscr{L}(\mathbf{H})$ for any fixed $\rho \in \mathscr{S}(\mathbf{H}$ as

$$\omega_\rho(\sigma) := tr(\rho\sigma) \in \mathbf{Q}_p(\mu) \quad \sigma \in \mathscr{L}(\mathbf{H}).$$

That is, ω_ρ is a *normalized involution preserving linear functional* on $\mathscr{L}(\mathbf{H})$.

- Therefore the sequence $\{\omega_\rho(M_i) = tr(\rho M_i)\}_{i \in I}$ serves as p-adic probability distribution.

8.2 p-Adically Quantumizing a Game

We summarize by describing a procedure to p-adically quantumize a strategic game: we first leverage the salient features of the p-adic probability in the following sense:

- We implement the states of an n-dimensional p-adic quantum system, associated with an n-pure strategies game, say, the 2-strategy game of the Battle of the Sexes $S = \{O, F\}$, representing the options of an Opera outing and an evening football game. Initially, we solve the game to identify two pure Nash Equilibria (PNE) at (O, O) and (F, F), which is not quite helpful for the wife and husband players decision-making.
- Then the players randomize the O and F strategies via a classical probability distribution, which yields a single Mixed Nash Equilibrium (MNE) at $\sigma^* = (\sigma_w, \sigma_h)$, with $\sigma_w^* = (p^*, 1 - p^*)$, and $\sigma_h^* = (q^*, 1 - q^*)$. Having now three Nash Equilibria is not ideal for the married couple decision-making, which prompts a need of a better coordination; being in a marriage could stand for some type of correlation between the couple's pure strategies.
- A strong, non-classical correlation might be required to generate an improved Nash Equilibrium. The pure strategies are therefore quantumized as described above, leading to the superposition and entanglement of the pure strategies; the entangled strategies produce a much better Nash Equilibrium, one that is Pareto-Optimal.
- The efficiency of this unique Pareto-Optimal Nash Equilibrium may be greatly improved by leveraging the distinguishing features of p-adic analysis:
- Expressing the payoffs in the p-adic numbers system $\mathbb{Q}_p$ to allow the ultranorm to estimate, for example, the distance between the payoffs, in particular following the entanglement of the strategies.
- The evolution of the strategies may have multiple layers which can be analyzed through the hierarchical structure of the p-adic analysis
- This process of p-adically quantumizing this particular game of the Battle of Sexes may therefore result in a single Pareto-Optimal Nash Equilibrium with high efficiency.

9 A Socio-cultural-Economical Game Theoretic Model

We present an example of how an advanced game-theoretic approach, such as a p-adic quantum game-theoretic one, can be used in the least expected area: the socio-cultural-economical dynamic of a society: modeling the strategic interactions of American using ten (10) pure strategies representing the ten core values in the American society

9.1 Game Theoretical Approach to America Cultural Evolution

Culture[26] consists of preference distribution along with equilibrium behavior distribution, but also individuals in any given culture are multidimensional. Immigration has been a major driving and defining factor of American culture. The behavior and choices of immigrants are influenced by many factors, e.g., home country traditions, in interaction with other American, while the preferences and equilibrium behaviors are shaped by *socialization and self-persuasion* and vary continuously, impacted by cross-cultural contacts, leading as a result, to some type of cultural hybridization. Discrepancies between the ideal and actual choices increase with time and fuel discontent. One would like to make choices in agreement with personal preferences, but the gains and payoffs are dependent upon the degree of choice coordination. The interactions, while seemingly random, are enhanced by factors such as common language, shared symbols and meanings, communication rules, down to culinary habits, family size.

America, as a social entity, has been experiencing rapidly changing demographics, with varying cultural norms and values. An American culture distinctly emerged as a compromise, a *modus vivendi* and *modus operandi* out of multiple early immigrant cultures, behaviors and preferences. These cross-cultural interactions generate a single hybrid culture, with its own peculiar language and rules of laws. The basic beliefs, assumptions and values by which most American live could be comprehensively described by some ten core values, we here recall, deeply ingrained in most Americans.

1. **Individualism and Privacy** (A_1): Each person is seen as unique, special and independent. Such belief put a premium on individual initiative, expression, orientation and values privacy. Other cultures emphasize group orientation, conformity as a way to societal harmony. American believe they control their own destiny. Other cultures' focus is on extended family, and its corollary of loyalty and responsibility to family, with age given status and respect.
2. **Equality/Egalitarianism** (A_2): Each person expects to be treat equally, with a minimum sense of hierarchy, leading to a directness in relations with others, informal sense of self and space. One by-product is to allow the challenging of authority as opposed to respect for authority and social order at all cost prevalent in other cultures. Gender equity rather than different roles for men and women. In other countries people seem to draw a sense of security and certainty from

[26] Some view culture as a vehicle for providing generally accepted solutions to problems. Others define it as part of the knowledge owned by a substantial segment of a group but not necessarily by the general population.

Class and Authority, Rank and Status, considering it reassuring to know, from birth, who you are and where you fit in the complex social system.

3. **Materialism** (A_3): Each person enjoys the right to be "well off" and to "pursuit material happiness". People are often judged by their possessions. An undesirable by-product is the well-documented *American Greed*. It is a higher priority to obtain, maintain and protect material objects rather than developing and enjoying interpersonal relationships for instance.
4. **Science and Technology** (A_4): This value is seen as the driving force for changes and the primary source of goods. Scientific and technological approaches are highly valued, and lead to a problem-solving focus, with a mental process and learning style that are linear, logical and sequential.
5. **Progress and Change** (A_5): This leads to a great national optimism and the so-called "Manifest Destiny": nothing is impossible and greatness is ours, as opposed to just accepting life's difficulties. Other societies value stability, tradition, continuity and rich and ancient heritage, seeing change as disruptive and potentially destructive.
6. **Work and Leisure** (A_6): This refers to a strong work ethic, and work for its intrinsic values and as the basis of recognition and power. Idleness is seen as a threat to society while leisure should be a reward for hard work and individual achievement. In other cultures, work is seen as a necessity of life, and rewards are based on seniority and relationships.
7. **Competition** (A_7): That is, the "Be always First" mentality, encouraging an aggressive and competitive nature. It has led to *Free Enterprise* as an economic system, in the belief that competition fosters rapid progress. Cooperation is instead promoted in other cultures.
8. **Mobility** (A_8): This refers to a vertical social, economic and physical mobility, in a society with people on the move to better self.
9. **Volunteerism** (A_9): Philanthropy is highly valued as a personal choice not a communal expectation, involving associations or denominations rather than kin-groups as in other cultures.
10. **Action and Achievement Oriented** (A_{10}): That is, a practical mind set with emphasis on getting things done, a focus of function and pragmatism and a tendency to be brief and business-like through an explicit and direct communication style, with emphasis on content and meaning found in the choice of words. In other cultures the emphasis is on context, meaning is found around the words, and communication is implicit and indirect. Formal handshakes rather than hugs and bows! *Dress for success* rather than as a sign of position, wealth, prestige or religious rules.

Some of these core values, in particular Individualism, Materialism, and Competition, are perceived sometimes by individuals from other cultural identities, at least during an initial period, with a negative and derogatory connotation. However, a complete cultural integration and economic success in America is for newcomers to be familiarized as quickly and efficient as possible with these values. We consider the cultural evolution strategies to be associated with these ten core values, and we define a probability distribution, Archimedean and Non-Archimedean, for each strategy $A_i, i = 1, \ldots, 10$. We are interested in determining the conditions for social stability and progress, and in designing a mathematical framework to understand the American social dynamics, by analyzing the interactive decision-making of all American in relation these ten core values. See Toni [78].

All the above aspects of game theoretic approach are applied to decipher the American socio-cultural evolution:

- Americans as players randomize the 10 pure strategies $\{A_1, \ldots, A_{10}\}$, assigning in their daily and random mutual interactions a probability to each one of the core values, leading to an evolutionary matrix game, as described above. The main Nash Equilibrium, represented by the American Constitution, is indeed a mixed Nash Equilibrium co-existing with many local Nash Equilibria, e.g., resulting from strategic interactions e.g., along party lines, socio-economics, religious, ethnic identities, gender, age, etc. These mixed sub-Nash equilibria are enclosed by limit cycles, we previously called Nash Limit Cycles, i.e., self-sustained oscillations in decision-making by American individuals, before converging to or diverging from the equilibrium points.
- These 10 American pure strategies/core values may also be *quantumized*, realistically, using superposition of these values, as well as entanglement of these values. Indeed, these American values are correlated, strongly correlated as times, possibly with some non-classical correlation (entangled American core values). This may result in more efficient Nash Equilibrium, Pareto-Optimal, abstract representation of, e.g., a socially optimum American Constitution with necessary amendments for social stability and progress.
- p-adically quantumizing these strategies/core values will certainly lead to an even more optimum Nash Equilibrium leveraging the nuances and hierarchy of the payoff layers. This is indeed the topic of an upcoming research work.

10 Concluding Remarks

This comprehensive compendium on advances in game theory provides a deep insight into the directions of game theory methodologies permeating all aspects of human knowledge.

Machine Learning (ML) and Artificial Intelligence (AI) techniques and their unparalleled computational power are expanding at a lightning speed: OpenAI and its *ChatGPT*, the Chinese *DeepSeek* are the dominant ones. Algorithmic Game Theory has been aggressively pursuing the design of efficient algorithms for the computing of the Nash Equilibrium. Combining all these techniques with classical, quantum and non-Archimedean/p-adic Game Theory is opening new frontiers in interactive strategic decision-making and predictive analysis across all fields on human knowledge. We are indeed witnessing the emerging of a very powerful tool integrating learning, adaptation and complex decision-making under uncertainty to model and solve real-world problems and beyond: artificial general intelligence, life extension, uploading of consciousness or mind uploading, people and their environment (e.g., food security and biodiversity), and more importantly post-human mathematics (i.e., artificial mathematical creativity) [68]. Game Theory, Artificial Intelligence and Quantum Computing are rapidly converging to generate a profound impact on how we understand, address and solve complex strategic problems.

References

1. Adler, S. S. 1995. *Quaternionic quantum mechanics and quantum fields*. Oxford.
2. Aniello, P., S. Mancinin, and V. Parisi. 2022. *Trace class operators and states in p-adic quantum mechanics*. arXiv:2210.01566v2 [math-ph]
3. Aniello, P., S. Mancinin, and V. Parisi. 2023. A p-adic model of quantum states and the p-adic qubit. *Entropy* 25: 86.
4. Aumann, R. J. 1959. Acceptable points in general cooperative n-person games. In *Contributions to the theory of games*, Annals of mathematics studies, vol. 40, ed. Luce, R. D. and A. W. Tucker. Princeton, NJ: Princeton University Press, 287–324
5. Aumann, R. J. 1974. Subjectivity and correlation in randomized strategies. *Journal of Mathematical Economics* 1 (1): 67–96.
6. Aumann, R. J., and M. Maschler. 1985. Game theoretic analysis of a bankruptcy problem for the Talmud. *Journal of Economic Theory* 36: 195–213.
7. Axelrod, R. 1984. *The evolution of cooperation*. New York: Basic Books.
8. Balkenborg, D., and K. Schlag. 2000. Evolutionarily stable sets. *International Journal of Game Theory* 29: 571–595.
9. Banach, S. 1922. Sur les opérations dans les ensembles abstraits et leur application aux équations intégrales. *Fundamenta Mathematicae* 3: 133–181.
10. Bellman, R. 1957. *Dynamic programming*. Princeton: Princeton University Press.
11. Benaim, M., and M. W. Hirsch. 1999. Mixed equilibria and dynamical systems arising from repeated games. *Games and Economic Behavior* 29: 36–72.
12. Berlekamp, E., J. H. Conway, and R.K. Guy. 1982. *Winning ways for your mathematical plays*, vol. 4. AK Peters/CRC Recreational Mathematics Series.
13. Binmore, K. 2007. *Playing for real: a text on game theory*. Oxford University Press.
14. Bisin, A., and T. Verdier. 2001. The economics of cultural transmission and the dynamics of preferences. *Journal of Economic Theory* 97: 298–319.
15. Bomze, I. M. 1983. Lotka-Volterra equation and relicator dynamics: A two-dimensional classification. *Biological Cybernetics* 48: 201–211.
16. Bowles, S. S., and H. Gintis. 2004. *The moral economy of communities: Adapting to market liberalism and the new economy*. Springer.
17. Brafman, R. I., and M. Tennenholtz. 2003. Rational and efficient interaction in multi-agent systems. In *Proceedings of the international joint conference on artificial intelligence (IJCAI)*.
18. Brams, S. J. *Biblical games: A strategic analysis of stories in the old testament*. Cambridge: MIT Press.
19. Brouwer, L. E. J. 1911. Über Abbildungen von Mannigfaltigkeiten. *Mathematische Annalen* 71: 97–115.
20. Carmona, R. 2020. *Application of mean field games in financial engineering and econmic theory*. arxiv:2012.05237
21. Cournot, A. 1838. *Recherches sur les principes mathématiques de la théorie des richesses*. New York: Macmillan.
22. Das, Sowmitra. 2023. *Quantumizing classical games: An introduction to quantum game theory*. arXiv:2305.00368v1 [quant-ph]

23. Debreu, D. 1952. A social equilibrium existence theorem. *Proceedings of the National Academy of Sciences* 38: 886–893.
24. Dixon, John D. 1982. Exact solution of linear equations using p-adic expansion. *Numerische Mathematik* 40: 137–141.
25. Djehiche, B., A. Tcheukam, and H. Tembine. 2017. Mean-field-type games in engineering. *AIMS Electronics and Electrical Engineering* 1 (1): 18–73.
26. Dockner, E., S. Jorgensen, N. Van Long, and G. Sorger. 2000. *Differential games in economics and management science*. Cambridge, UK: Cambridge University Press.
27. Eisert, J., and M. Wilkesn. 2000. Quantum games. *Journal of Modern Optics* 47 (14–15): 2543–2556.
28. Eisert, J., M. Wilkens, and M. Lewenstein. 1999. Quantum games and quantum strategies. *Physical Review Letters* 83: 3077–3080.
29. Fan, K. 1952. Fixed point and minimax theorems in locally convex topological linear spaces. *Proceedings of the National Academy of Sciences* 38: 121–126.
30. Farell, J., and R. Ware. 1989. Evolutionary Stability in the repeated prisoner's dilemma. *Theoretical Population Biology* 36: 161–166.
31. Fisher, R. A. 1922. On the dominance ratio. *Proceedings of the Royal Society of Edinburgh* 42: 321–341.
32. Fudenberg, D., and J. Tirole. *Game theory*. MIT Press.
33. Gilboa, I., and A. Matsui. 1991. Social stability and equilibrium. *Econometrica* 59: 859–867.
34. Glicksberg, I. L. 1952. A further generalization of the Kakutani fixed point theorem with applications to Nash equilibrium points. *Proceedings of the National Academy of Sciences* 38: 170–174.
35. Gouvê a, F. Q. 1997. *p-adic numbers, an introduction*. Berlin, Germany: Springer.
36. Gregory, R. T. 1980. *Error-free computation*. Huntington: Krieger.
37. Harper, D. G. 1982. Competitive foraging in mallards: "Ideal free" ducks. *Animal Behaviour* 30 (2): 575–584.
38. Harsanyi, J. C., and R. Selten, R. 2003. *A general theory of equilibrium selection in games*. Cambridge, MA: MIT Press.
39. Hofbauer, J. 1981. On the occurrence of limit cycles in the Lotka-Volterra differential equation. *Journal of Nonlinear Analysis* 5: 1003–1007.
40. Hofbauer, J., and K. Sigmund. 1998. *Evolutionary games and replicator dynamics*. Cambridge: Cambridge University Press.
41. Hofbauer, J., and K. Sigmund. 2003. Evolutionary game dynamics. *Bulletin of the American Mathematical Society* 40: 479–519.
42. Hofbauer, J., P. Schuster, and K. Sigmund. 1979. A note on evolutionary stable strategies and game dynamics. *Journal of Theoretical Biology* 81: 609–612.
43. Iqbal, A., and A. Toor. 2001. Evolutionary stable strategies in quantum games. *Physics Letters A* 280 (5–6): 249–256.
44. Kakutani, S. 1941. A generalization of Brouwer's fixed-point theorem. *Duke Mathematical Journal* 8 (3): 457–459.
45. Katok, S. 2007. *P-adic analysis compared with real*. AMS Student Math Library, vol. 37. American Mathematical Society, Providence. RI.
46. Khan, F. S., and S. Phoenix. 2013. Gaming the quantum. *Quantum Information and Computation* 13: 231–244.
47. Khrennikov, A. 1997. *Non-archimedean analysis: Quantum paradoxes, dynamical systems and biological models*, 1st ed. Alphen aan den Rijn, the Netherlands: Kluwer Academic Publishers.
48. Khrennikov, A. 1993. p-adic probability theory and its applications. The principle of statistical stabilization of frequencies. *Theoretical and Mathematical Physics* 97: 1340–1348.
49. Koblitz, N. 1984. *p-adic numbers, p-adic analysis, and zeta-functions*. Springer.
50. Krabs, W., and S. Pickl. 2010. *Dynamical systems, stability, controllability and chaotic behavior*. Berlin, Heidelberg: Springer.
51. Krishnamurty, E. V. 1977. Matrix processors using p-adic arithmetic for exact linear computations. *IEEE Transactions on Computers* 26 (7): 633–639.

52. Lasry, M., and P. L. Lions. 2007. Mean field games. *Japanese Journal of Mathematics* 2 (1): 229–260.
53. Luce, R. D., and H. Raiffa. 1958. *Games and decisions: Introduction and critical survey*. New York: Wiley.
54. Marinatto, L., and T. Weber. 2000. A quantum approach to static games of complete information. *Physics Letters A* 272 (5): 291–303.
55. Meyer, D. A. 1999. Quantum strategies. *Physical Review Letters* 82: 1052–1055.
56. Nasar, S. 1998. *A beautiful mind*. London: Faber and Faber Ltd.
57. Nash, J. 1950. Equilibrium points in n-person games. *Proceedings of the National Academy of Sciences of the United States of America* 36: 48–49.
58. Nash, J. F. 1950. The bargaining problem. *Econometria* 18: 155–162.
59. Nash, J. 1951. Non-cooperative games. *Annals of Mathematics* 54 (2): 287–295.
60. Nash, J. F. 1953. Two-person cooperative games. *Econometria* 21: 128–140.
61. Neeman, A. 2014. *Non-Archimedean analysis*. Springer.
62. Nisan, N., Roughgarden, T. , Tardos, E., Vazirani, V. 2007. *Algorithmic game thoery.*Cambridge University Press, Cambridge UK
63. Nowak, M. A. 2006. *Evolutionary dynamics*. Cambridge, USA: Harvard University Press.
64. Packel, E. 2006. *The mathematics of games and gambling*. The Mathematical Association of America, vol. 28
65. Robert, A. M. 2000. *A course in p-adic analysis*. Springer.
66. Rosenthal, R. W. 1973. *A class of games possessing pure-strategy Nash Equilibrium*. Springer.
67. Roughgarten, T. 2016. *Twenty lectures on algorithmic game theory*. Cambridge University Press.
68. Ruelle, D. 2013. *Post-human mathematics*. arXiv:1308.4678 [Math.Ho]. https://doi.org/10.48550/arXiv.1378.4
69. Ruffa, A., M. Jandron, and B. Toni. 2016. Parallelized solution of banded linear systems with an introduction to p-adic computation. In *Mathematical sciences with multidisciplinary applications*, vol. 157. Springer PROMS 431–464.
70. Sandholm, W. H. 2001. Preference evolution, two-speed dynamics, and rapid social change. *Review of Economic Dynamics* 4: 637–679.
71. Schauder, J. 1930. Über lineare, vollständig kontinuierliche Operatoren. *Studia Mathematica* 2: 1–6.
72. Schecter, S. 2012. *How the talmud divides an estate among creditors*. Toni et al. (eds) Springer PROMS 24:29–42
73. Schuster, P., and K. Sigmund. 1983. Replicator dynamics. *Journal of Theoretical Biology* 100: 533–538.
74. Shapley, L. S. 1967. on balanced sets and cones. *Naval Research Logistics Quarterly* 14: 453–460.
75. Shoham, Y., and K. Leyton-Brown. 2009. *Multiagent systems: Algorithms, Game-Theoretic and Logical Foundation*. Cambridge University Press.
76. Smith, M. J. 1982. *Evolution and the theory of games*. Cambridge: Cambridge University Press.
77. Taylor, P. D., and L. B. Jonker. 1978. Evolutionary stable strategies and game dynamics. *Mathematical Biosciences* 40: 145–156.
78. Toni, B. 2017. Nash limit cycles: A game-theoretical analysis of cultural integration in America. In *New trends and advanced methods in interdisciplinary mathematical sciences*. Springer STEAM-H, 321–356.
79. Tzu, Sun. 1988. *The art of war* [Transl. by Thomas Cleary]. Boston: Shambala.
80. Vladomirov, V. S., I. V. Volovich, and E. I. Zelenov. 1994. *p-adic analysis and mathematical physics*. Singapore: World Scientific.
81. von Neumann, J., and O. Morgenstern. 1947. *Theory of games and economic behaviour*. Princeton UP
82. Webb, J. 2007. *Game theory: Decisions, interaction and evolution*. Springer Nature.
83. Weibull, J. 1995. *Evolutionary game theory*. Cambridge, USA: MIT Press.

84. Zeeman, E. 1980. Population dynamics from game theory. In *Global theory of dynamical systems*. Lecture Notes in Mathematics, vol. 819. New York: Springer.
85. Zeeman, E. 1981. Dynamics of the evolution of animal conflicts. *Journal of Theoretical Biology* 89: 248–270.
86. Zermelo, E. 1913. Uber eine anwendung der mengenlehre auf die theorie des schachspiels. In *Proceedings fifth international congress of mathematicians*, vol, 2, 501–504.

Information-Delay-Induced Hopf Bifurcations in Evolutionary Game Dynamics

Thomas A. Wettergren

Abstract Evolutionary games describe the repeated random interactions of individuals with others in a large population. In these interactions, individuals have a choice between strategies that enable them to pursue a group interest (cooperate) or pursue a self-interest (defect). When the individuals have a capacity to change this selection of strategy based on their interactions, and the decisions are based on performance that is influenced by the choice of others, then the evolution has been shown to lead to a stable percentage of cooperators over time. The dynamics of such an evolution of strategies can be described by individuals adjusting their strategy either (i) based on comparison of their performance to that of others (imitation update rules) or (ii) based on assessing their performance against a goal (aspiration update rules). In either case, these dynamics are typically modeled as an instantaneous reaction to the observations made by the individuals. However, in real physical and social systems, there are often non-negligible time delays between the observations and the ability of an individual to adapt. When these time delays become large enough, they can create an instability in the dynamics of the system strategies pursued by the individuals (as measured by the percentage of individuals following the cooperation strategy). We investigate the initiation of this instability via a Hopf bifurcation in the time delay parameter. In addition to explaining the transition, this analysis also yields guidance on how much time delay is tolerable in order to guarantee convergence of the group behavior. Both analytical and numerical results are presented to illustrate the phenomenon.

Keywords Evolutionary game · Dynamical system · Hopf bifurcation · Stability · Performance analysis

T. A. Wettergren (✉)
Naval Undersea Warfare Center, Newport, RI, USA
e-mail: t.a.wettergren@ieee.org

E. A. Pimentel and B. Toni (eds.), *Differential and Algorithmic Intelligent Game Theory*, STEAM-H: Science, Technology, Engineering, Agriculture, Mathematics & Health,
https://doi.org/10.1007/978-3-031-97733-6_2

1 Introduction

Noncooperative game theory examines the impact of various choices taken by players who are impacted by the choices made by others in situations where these choices must be made without prior knowledge or consultation of the choices made by the others. When a large population of individuals interacts with one another through recurring noncooperative game interactions, the resulting process is an *evolutionary game* [7, 15, 19]. These evolutionary games have been used to model the propagation of species over multiple generations as well as scenarios of multiple individuals interacting repeatedly to accomplish a complex task. In all such situations, the models become a Markov-style process of repeated interactions between subsets of the population, wherein the strategies pursued by individuals (in terms of choices that are made for each interaction) evolve over time based upon past experiences.

Due to the evolving temporal nature of evolutionary games, the changes in strategies that are pursued by individuals follow various dynamics. Some of the most basic dynamics involve imitation behavior, whereby individuals change their strategy in the future to a better performing strategy that they've observed through an interaction. In this case individuals benefit from interacting regularly with a variety of other players. Because of this, the imitation behavior is mostly observed in systems with large and well-mixed populations of individuals. Another basic dynamic comes from an aspirational process, whereby individuals examine their reward from a given interaction and change their strategy if the amount of the reward is below some aspirational goal. The development of a basic aspiration-level goal has been well-studied [11] and also has been extended to where different players have different goals [12]. Whether using imitation or aspiration based dynamics, the dynamic behavior creates a learning mechanism in those situations when the system dynamics stabilize to a desired end-state. The dynamic aspect also creates a feedback mechanism such that the players can naturally adjust to changes in the environment [2].

The stability properties of these evolutionary game dynamics are subject to practical limitations, one of which is the time delay that it may take for one individual to observe the behavior of other individuals. We thus concern ourselves with the impact of time delays on the stability properties of evolutionary games. Tao and Wang [17] were the first to examine a time delay in a two-player game and showed that it can affect the stability properties. These delays can occur in two forms [1], referred to as the social-type delay (based on payoffs from the past) and the biological-type delay (based on offspring from parents in the past). We focus our study on the social-type delay, as that is consistent with delays due to communications in engineered systems.

It has been previously shown that the basic replicator dynamics for games with imitation-style behaviors have transitions from stable equilibria to oscillations as the time delay is increased beyond a critical value [10]. This type of delay-induced transition from stability has also been shown in systems which have added effects of mutation [9], as well is in systems with finite population sizes [21]. The analysis of the stability properties of evolutionary games has been approached with a variety of mathematical techniques, including variants of Lyapunov analysis applied to networks [5] and the D-composition method [3]. We apply a stability analysis that examines the Hopf bifurcations in the characteristic equations about known equilibrium points, as shown in Wesson et al. [20].

In this chapter, we illustrate the effect of delays on the stability of evolutionary game dynamics for games with both imitation and aspiration dynamics. It is shown that large enough delays can cause a loss of stability of the equilibrium points that were stable in the absence of the delay. By performing a Hopf bifurcation analysis, we both demonstrate the mechanism by which the destabilization occurs and also derive expressions for the specific amount of delay at which it occurs. Using this analysis, systems which are designed with this type of game-theoretic learning mechanism can be developed with delays that are below that threshold, thus guaranteeing the stability property of the equilibrium, which is important in applications.

1.1 Models of Two-Player Game Interactions

Evolutionary games are repeated plays of a strategy that are made among a population whose members employ variations of that strategy [19]. We refer to the individuals from the population as the *players* in the game. In the most basic evolutionary games, the interaction between players only occurs pairwise (the two-player games) as individual players encounter other players. In these interactions each player can choose to either behave cooperatively with the other player in the interest of the group, or avoid the cooperation and pursue a self-interest (referred to as defecting). These basic cooperate and defect behaviors represent the *strategies* available for each player to play during each interaction. If the payoffs that are achieved by each player only depend on the strategies employed and not on the specific individuals, then the game is *symmetric*. A general model of the two-player interactions for a symmetric game is given by the payoff matrix Π:

$$\Pi : \quad \begin{array}{c|c|c} & C & D \\ \hline C & R & S \\ \hline D & T & P \end{array} \tag{1}$$

where the component π_{ij} of Π is the payoff rewarded to a player who plays strategy $i \in \{C, D\}$ against an opponent who plays strategy $j \in \{C, D\}$ (i.e. a player playing strategy C against an opponent playing strategy D receives a payoff of $\pi_{CD} = S$, whereas the opponent receives a payoff of $\pi_{DC} = T$).

It is well-known that the relative strengths of the four components of the payoff matrix lead to four fundamental two-player game forms [18]. When $T > R > P > S$, the game is of the form of a Prisoner's Dilemma game. When $R > T \geq P > S$, it is a Stag Hunt game. When $T > R > S > P$, it is a Hawk–Dove game (which is also called the Chicken or Snowdrift game). Finally, when $R > S > T > P$, it is a Harmony game (note that a fifth type of game, the Deadlock game, has the reverse ordering of the Harmony game, yet we consider them to be of the same form since they are identical when the roles of cooperate and defect are switched). As repeated interactions between randomly paired individuals occur over time, the game evolves and the proportion of players choosing each strategy changes over time. By treating the fraction of players following each strategy as state variables, the dynamics of the evolution of the population's overall strategy can be observed as a dynamical system on the state variables. The study of the long-term behavior of these dynamics gives insight into whether the strategies pursued in a game approach a stable equilibrium or diverge into some other (typically cyclical) behavior.

In the absence of time delays, the Prisoner's Dilemma game has two equilibria that correspond to the two pure strategies (either always cooperate or always defect). The Stag Hunt game shares those two equilibria, but adds an additional mixed equilibrium, whereby some players cooperate while others defect. This mixed equilibrium is not observed in the evolutionary dynamics as it is always unstable. The Hawk–Dove game also has three equilibria; however, under certain combinations of payoff matrix parameters, the mixed equilibrium is known to be stable. The Harmony game has a single stable equilibrium which occurs at the pure strategy where both players cooperate. Since we are interested in non-trivial mixed equilibria, we focus our study on the Hawk–Dove game to illustrate the impacts of time delays on its mixed strategy equilibrium. In particular, we show conditions under which the equilibrium destabilizes under the time delay via a Hopf bifurcation, and also demonstrate how the inclusion of additional dynamics (due to slight modifications in the game-play) can change the critical time delay at which the Hopf bifurcation occurs.

1.2 Hawk–Dove Game Payoffs

We consider a Hawk–Dove game corresponding to competition over a shared resource, to which we assign a reward value $v > 0$. The competition may represent determining how much of a role to take in performing some task, or it may be a biological struggle for obtaining a portion of available nutrients, or it may be any other resource conflict between individuals of the same kind. In any case, the shared resource becomes contested between interacting players. Typically this interaction occurs between randomly paired individuals, wherein each competition is an increment of time, and the repeated time increments with interactions between varying individual players represents the evolutionary game.

Each player in the pairwise interactions in this game has a choice between two strategies: they can passively accept their share of the resource without conflict (the Dove strategy, which we refer to as *cooperation*) or they can actively attempt to claim a larger portion and fight if needed (the Hawk strategy, which we refer to as *defection*). If both players cooperate (playing the Dove strategy), then there is no conflict and the resource is shared evenly, giving each player an equal share of the reward for a payoff of $v/2$. If a player defects (plays Hawk strategy) when the opponent cooperates (plays Dove strategy), then the defector receives a payoff of the entire reward value v and the cooperator receives nothing. On the other hand if both players defect, the resulting conflict comes at a cost of c, leading to a payoff for each of $(v - c)/2$. We assume that this cost is larger than the reward value, such that $c > v$, since that creates a dilemma whereby cooperation is desirable over defection when the opponent defects. Note that in cases where $c < v$, the pure defection strategy is a Nash equilibrium, and instead of a Hawk–Dove game, the strategy space becomes a Prisoner's Dilemma game.

The payoff matrix for the two-player Hawk–Dove game as described above is given as

Π : (2)

	C	D
C	$v/2$	0
D	v	$(v - c)/2$

As the strategy preferences (and hence the qualitative evolutionary dynamics) for two-player games with constant payoffs are invariant under positive affine transformations of the payoff matrix, we make such a transformation to transform the payoff matrix into a more convenient form for analysis. See Sect. 1.3.3 in [19] for the invariance properties of strategy preferences under affine transformations. Applying the following positive affine transformation to the payoff matrix:

$$\Pi \quad \mapsto \quad (2/c)\Pi + (c - v)/c \tag{3}$$

and introducing $\rho = v/c$ as the benefit-to-cost parameter leads to a single parameter form of the payoff matrix

Π : (4)

	C	D
C	1	$1 - \rho$
D	$1 + \rho$	0

where we note that $0 < \rho < 1$ since $c > v > 0$, by assumption. The benefit-to-cost parameter ρ represents the ratio of the available reward value relative to the cost of cooperation. Thus, larger values of ρ imply an efficiency in cooperation (and conversely smaller values imply less efficiency). Specifically, this means an increase in the value of ρ can be either due to a cheaper cost for the same reward, or due to a greater reward for the same cost. In terms of the general game parameters, the payoff matrix Π in (4) yields $\{R, S, T, P\} = \{1, (1 - \rho), (1 + \rho), 0\}$, such that $T > R > S > P$, which is the condition mentioned above for a Hawk–Dove game. The payoff matrix form (4) is the form used throughout this chapter. We reiterate that the strategy preferences within this strategy space are the same as that for payoff matrix (2), due to the invariance property of payoff matrices under positive affine transformations.

2 The Imitation Game Dynamics

In many types of systems where large groups of individuals repeatedly interact with one another, the individuals adjust their strategy (to either cooperate or defect) based on how they perform relative to others that they encounter. We model this process using evolutionary game theory by considering repeated pairwise interactions between players who are randomly chosen from the population. This is commonly referred to as the *imitation game*, as players may adjust their strategy through imitation of the strategies that are being pursued by others.

The imitation game assumes that players occasionally choose another player at random and then adopt that other player's strategy with a probability depending on the relative payoffs of the two players [19]. The corresponding *imitation dynamics* refers to the resulting dynamical behavior due to the evolution of the fraction of players performing each strategy over time. For example, consider a player in an evolutionary game with cooperate/defect strategies who chooses to cooperate, and let $x(t)$ be the fraction of such cooperating players in the population at time t. Let $\pi_C(t)$ represent the expected payoff for a cooperating player from the population at time t, and $\pi_D(t)$ be that for a defecting player. From the game matrix form (1), the expected payoffs $\pi_C(t)$ and $\pi_D(t)$ are written in terms of $x(t)$ as

$$\pi_C(x(t)) = Rx(t) + S(1 - x(t)) = S + (R - S)x(t) \tag{5}$$

and

$$\pi_D(x(t)) = Tx(t) + P(1 - x(t)) = P + (T - P)x(t) \tag{6}$$

respectively. These are the expected values of the rewards that could be achieved by an individual performing each of the strategies of C or D, based on the assumed behavior of others, which is given by the assumed fraction of cooperators $x(t)$ and the corresponding assumed fraction of defectors $(1 - x(t))$.

For the imitation game, the rate of change of $x(t)$ is given by an input-output model of the form [8]

$$\frac{dx(t)}{dt} = x(t)(1 - x(t))\,[f_{CD}(x(t)) - f_{DC}(x(t))] \tag{7}$$

where f_{CD} is the rate at which a defecting player switches to cooperation, and conversely for f_{DC}. If we assume that the rate functions f only depend on the payoffs of the players, then we have that

$$f_{CD}(x(t)) = \psi(\pi_C(x(t)), \pi_D(x(t))) \tag{8}$$

and

$$f_{DC}(x(t)) = \psi(\pi_D(x(t)), \pi_C(x(t))) \tag{9}$$

where $\psi(u, v)$ is a function that defines the *imitation rule* for the game. A common assumption is to assume the imitation rule function $\psi(u, v)$ only depends on the payoff difference, such that $\psi(u, v) = \psi(u - v)$ [7].

The simplest form of an imitation rule is to have a player imitate the strategy of the other player whenever that player has a better payoff than oneself. In that case, the imitation rule function $\psi(u, v)$ is given by

$$\psi(u, v) = \begin{cases} 0, & u - v \leq 0 \\ 1, & u - v > 0 \end{cases} \tag{10}$$

However, the discontinuity created by this function makes analysis difficult. If we instead assume that the imitation rule function follows the *proportional imitation rule* in which players probabilistically imitate strategies that yield a higher payoff, where the probability of switching is proportional to the expected gain, then the imitation rule function takes the form [8]

$$\psi(u, v) = \begin{cases} 0, & u - v \leq 0 \\ \alpha(u - v), & u - v > 0 \end{cases} \tag{11}$$

for some constant $\alpha > 0$. Applying this proportional imitation rule (with $\alpha = 1$) to the basic imitation dynamics of Eq. (7) yields the following form of imitation dynamics:

$$\frac{dx(t)}{dt} = x(t)(1 - x(t)) \left[\pi_C(t) - \pi_D(t)\right] \tag{12}$$

It has been noted by Hofbauer and Sigmund [7, 8] that this formulation reduces the imitation dynamics to the well-known replicator dynamics for evolutionary games that mimic the process of natural selection over many generations of reproduction [19]. Thus, the replicator dynamics that are well-known in ecological games are representative of the imitation game dynamics for the special case of an imitation game with a probabilistic switching of strategies in which the probability is proportional to the expected payoff gain for making the switch.

2.1 Equilibria of the Imitation Game Dynamics

We are interested in group situations where there is a delay in the ability of players to receive information about the group, which we refer to as an *information delay*. This information delay corresponds to the type of time delay that Alboszta and Miekisz refer to as a *social-type time delay* [1]. Such delays reflect a limitation in the ability of individual players to quickly receive and assimilate information about the aggregate state of a large group, which is a common problem in applications. When a constant social-type time delay of τ is included in the imitation game dynamics described by

Eq. (12), it affects each player's knowledge of the assumed fractions of cooperators and defectors, thus replacing $x(t) \to x(t-\tau)$ in Eqs. (5) and (6), yielding

$$\pi_C^d(t) = S + (R - S)x(t - \tau) \tag{13}$$

and

$$\pi_D^d(t) = P + (T - P)x(t - \tau) \tag{14}$$

as delayed forms of the expected payoff for cooperators and defectors, respectively. This leads to an imitation game dynamics for the system with social-type delays of the form

$$\begin{aligned}\frac{dx(t)}{dt} &= x(t)\,(1 - x(t))\,\left[\pi_C^d(t) - \pi_D^d(t)\right] \\ &= x(t)\,(1 - x(t))\,[(S - P) + (R - S - T + P)x(t - \tau)]\end{aligned} \tag{15}$$

For the Hawk–Dove game of interest with a payoff matrix of the form given by (4), the delayed imitation game dynamics are given as

$$\frac{dx(t)}{dt} = x(t)\,(1 - x(t))\,[(1 - \rho) - x(t - \tau)] \tag{16}$$

This game clearly has three equilibria at $x = \{0, 1, (1 - \rho)\}$ where the pure strategy equilibria at $x = 0$ and $x = 1$ correspond to all players performing identically and the equilibrum at $x = x^* = (1 - \rho)$ is a mixed strategy equilibrium with some players performing each of the two pure strategies. We also note that the expected average payoff of the population $\bar{\pi}(t)$ for this Hawk–Dove imitation game takes the form

$$\begin{aligned}\bar{\pi}(t) &= x(t)\,\pi_C^d(t) + (1 - x(t))\,\pi_D^d(t) \\ &= (1 - \rho)x(t) + (1 + \rho)x(t - \tau) - x(t)x(t - \tau)\end{aligned} \tag{17}$$

At the mixed strategy equilibrium value of $x^* = (1 - \rho)$, the expected payoff becomes

$$\bar{\pi}^* = 1 - \rho^2 \tag{18}$$

Note that this equilibrium payoff value of $\bar{\pi}^*$ falls between the payoff that a cooperator would achieve against another cooperator (payoff of $R = 1$) and the payoff that a cooperator would achieve against a defector (payoff of $S = 1 - \rho$). In particular:

$$(1 - \rho) < (1 - \rho)(1 + \rho) = (1 - \rho^2) < 1 \qquad \forall\, \rho \in (0, 1) \tag{19}$$

Thus, the mixed strategy equilibrium payoff value in Eq. (18) can be thought of as a compromise value of what a cooperator might realistically expect to achieve for a payoff against uncertain opponents over time.

2.2 Stability of the Imitation Game Equilibria

In the absence of the time delay (i.e. when $\tau = 0$), the imitation game dynamics for the Hawk–Dove imitation game with the payoff form of (4) become

$$\frac{dx(t)}{dt} = x(t)\,(1 - x(t))\;[(1-\rho) - x(t)] \tag{20}$$

This equation has the same three equilbrium values at $x = \{0, 1, (1-\rho)\}$ as in the delayed form. The stability properties of each of these equibria can be assessed through standard linear characteristic analysis.

For the equilibrium at $x = 0$, the local linearized form of the dynamics are given by forming $\xi(t) = x(t)$, and maintaining only terms linear in $\xi(t)$ (for the small neighborhood assumption), to obtain

$$\frac{d\xi(t)}{dt} = (1-\rho)\xi(t) + O(|\xi(t)|^2) \tag{21}$$

Since $(1-\rho) > 0 \;\forall\; \rho \in (0, 1)$, this equilibrium is unstable for all parameter values of interest.

For the equilibrium at $x = 1$, the local linearized form of the dynamics are given by forming $\xi(t) = x(t) - 1$, and maintaining only terms linear in $\xi(t)$ (for the small neighborhood assumption), to obtain

$$\frac{d\xi(t)}{dt} = \rho\xi(t) + O(|\xi(t)|^2) \tag{22}$$

Since $\rho > 0 \;\forall\; \rho \in (0, 1)$, this equilibrium is also unstable for all parameter values of interest.

For the equilibrium at $x = (1-\rho)$, the local linearized form of the dynamics are given by forming $\xi(t) = x(t) - 1 + \rho$, and maintaining only terms linear in $\xi(t)$ (for the small neighborhood assumption), to obtain

$$\frac{d\xi(t)}{dt} = -\rho(1-\rho)\xi(t) + O(|\xi(t)|^2) \tag{23}$$

Since $-\rho < 0$ and $(1-\rho) > 0 \;\forall \rho \in (0, 1)$, this equilibrium is stable for all parameter values of interest. Hence, in the situation with no time delays, the only stable equilibrium is found at $x^* = (1-\rho)$, and thus all solutions will asymptotically approach this level of cooperation. This corresponds to a mixed strategy, in which players are neither all performing a cooperation strategy of the Dove (as in $x = 1$) nor all performing a defection strategy of the Hawk (as in $x = 0$). The mixed strategy equilibrium at x^* provides a stable balance between Hawk and Dove strategies. This equilibrium also is a Nash equilibrium, as all Lyapunov stable states of the replicator dynamics of an evolutionary game correspond to Nash equilibria [19], and we have previously shown that the imitation game dynamics reduce to the replicator dynamics for this situation. Since this equilibrium represents a balance of strategies at each time step, it can either be observed as a mixture of players with pure strategies (i.e. fraction x^* always behave as Dove and fraction $(1 - x^*)$ always behave as Hawk), or it can also be realized in practice as a mean value for a probability distribution on a randomized selection of strategies by individuals at each time step (i.e. an individual player behaves as Dove for $100 \times x^*$ percent of the time). In the presence of time delays, it is known that asymptotically stable equilibria such as this one can de-stabilize through a Hopf bifurcation [4]. In the next section, we examine when such a bifurcation may occur for this system.

2.3 Hopf Bifurcation Due to the Time Delay

We consider the delayed imitation game dynamic equation, repeated here:

$$\frac{dx(t)}{dt} = x(t)\ (1 - x(t))\ [(1 - \rho) - x(t - \tau)] \tag{24}$$

and we study the equilibrium at $x^* = (1 - \rho)$. To form a characteristic equation for this first-order delayed differential equation, we form two local variables in a neighborhood of the equilibrium point, one given by $\xi(t) = x(t) - (1 - \rho)$ and the second incorporating the delay as $\xi^d(t) = \xi(t - \tau) = x(t - \tau) - (1 - \rho)$. Replacing the terms $x(t)$ and $x(t - \tau)$ in the delayed imitation game dynamic equation with their local counterparts, we arrive at

$$\frac{d\xi(t)}{dt} = -\xi^d(t)\ (1 - \rho + \xi(t))\ (\rho - \xi(t)) \tag{25}$$

Keeping only terms that are linear in the local variables leads to the linearized form of the local imitation game dynamic equation as

$$\frac{d\xi(t)}{dt} = -\rho(1 - \rho)\xi^d(t) + O(|\xi(t)|^2) \tag{26}$$

where we have used the fact that $\xi(t)$ and $\xi^d(t)$ are of that same order of magnitude (that is, $O(|\xi(t)|) = O(|\xi^d(t)|)$ since they only differ by a constant time delay). Introducing the characteristic parameter λ through $\xi(t) = e^{\lambda t}$ and $\xi^d(t) = e^{\lambda(t-\tau)}$ creates a characteristic equation of the form

$$\lambda = -\rho(1 - \rho)\ e^{-\lambda\tau} \tag{27}$$

The stability properties of the equilibrium point at x^* are thus given by the location of the values of λ in the complex plane.

The analysis from the previous section shows that the equilibrium at x^* is asymptotically stable for all values of $\rho \in (0, 1)$ when $\tau = 0$. This is consistent with the characteristic Eq. (27), as it becomes $\lambda = -\rho(1-\rho) < 0$ when $\tau = 0$. As the time delay τ is increased from zero, there is a possibility of introducing a large enough time delay to throw the system into instability by de-stabilizing this equilibrium. This could occur through a Hopf bifurcation, where the values of λ cross into the right-hand-side of the complex plane. To find if and when this occurs, we look for the Hopf bifurcation by setting $\lambda = i\omega$ in the characteristic equation to correspond to the critical time delay of $\tau = \tau_c$ where the de-stabilization occurs. Making the substitution and collecting real and imaginary terms leads to the following two equations:

$$\text{Real:} \quad 0 = -\rho(1-\rho)\cos(\omega\tau_c) \;\Rightarrow\; \cos(\omega\tau_c) = 0 \tag{28}$$

and

$$\text{Imaginary:} \quad \omega = \rho(1-\rho)\sin(\omega\tau_c) \;\Rightarrow\; \sin(\omega\tau_c) = \frac{\omega}{\rho(1-\rho)} \tag{29}$$

By squaring Eqs. (28) and (29) and applying the Pythagorean identity, we obtain an equation for ω (without τ_c) that can be solved as

$$\omega = \rho(1-\rho) \tag{30}$$

Substituting the value of ω from Eq. (30) into Eq. (28) yields the value of τ_c, which is given by

$$\tau_c = \frac{\pi}{2\rho(1-\rho)} \tag{31}$$

Equation (31) gives the critical value of the time delay at which a Hopf bifurcation occurs. For values of $\tau < \tau_c$, the imitation game dynamics lead to an asymptotically stable value of x^*, whereas for values of $\tau > \tau_c$, the imitation game dynamics are unstable at x^*. Thus, the Hopf bifurcation point provides guidance on how much time delay a Hawk–Dove imitation game can tolerate to maintain stable imitation game dynamics.

2.4 Numerical Example

The dynamics of the Hawk–Dove imitation game with delays (as given in Eq. (16)) were simulated numerically to assess the system's long-term dynamic behavior. We are particularly interested in observing when and how the fraction of cooperators $x(t)$ leads to the mixed strategy Nash equilibrium of $x^* = 1 - \rho$. For all of the numerical simulations that are shown in this chapter, Matlab's dde23 solver for the numerical solution of delayed differential equations was utilized [16].

Consider a Hawk–Dove imitation game with delays and a benefit-to-cost parameter value of $\rho = 0.3$. The mixed strategy Nash equilibrium corresponding to this value of ρ occurs at $x^* = 1 - \rho = 0.7$. From the analysis of the preceding section, an expected Hopf bifurcation for the asymptotic stability of this mixed strategy equilibrium should occur at $\tau_c \approx 7.48$ (see Eq. (31)). Assume a small time delay of $\tau = 7.0$ (considered small because it is below the expected critical Hopf delay τ_c). The simulated dynamics for a Hawk–Dove imitation game beginning with an equal proportion of cooperators and defectors (i.e. $x(0) = 0.5$) is shown in Fig. 1, with a corresponding phase portrait drawn in Fig. 2. Clearly the numerical solution demonstrates asymptotic stability of the proportion of cooperators approaching the Nash equilibrium.

Next consider the same Hawk–Dove imitation game with a slightly larger time delay of $\tau = 8.0$. While this time delay is only slightly larger than the prior case, it

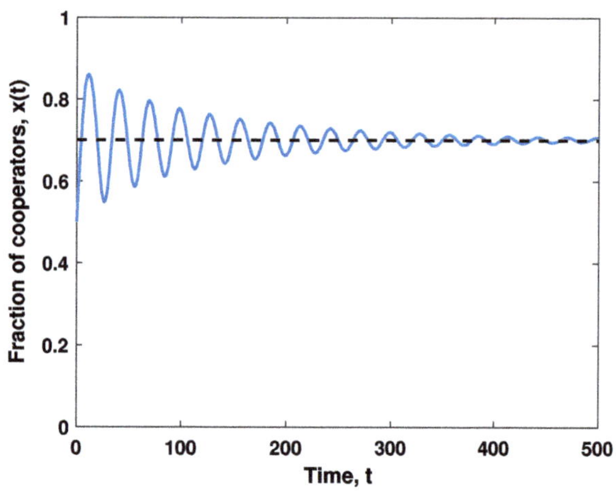

Fig. 1 Time series for the imitation game dynamics with $\rho = 0.3$ and a time delay of $\tau = 7.0$. The dashed line is the Nash equilibrium solution at $x^* = 0.7$

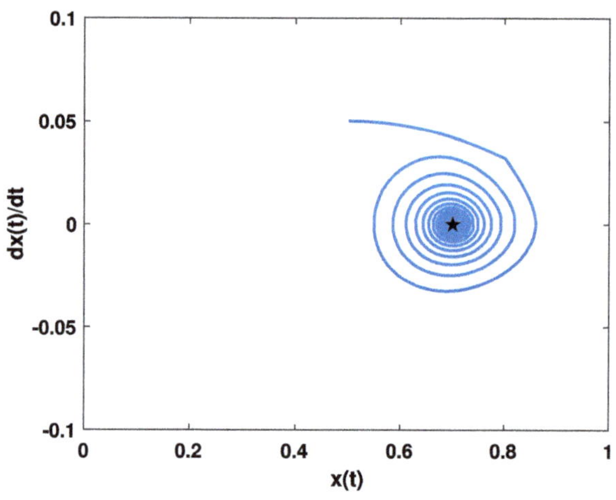

Fig. 2 Two-dimensional projection of the phase portrait for the example from Fig. 1. The phase portrait is drawn in the $x - \dot{x}$ planeTwo-dimensional projection of the phase portrait for the example from Fig. 1. The phase portrait is drawn in the $x - \dot{x}$ plane

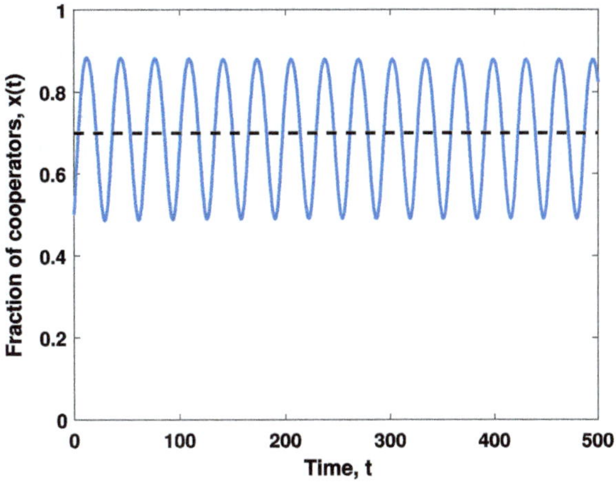

Fig. 3 Time series for the imitation game dynamics with $\rho = 0.3$ and a time delay of $\tau = 8.0$. The dashed line is the Nash equilibrium solution at $x^* = 0.7$

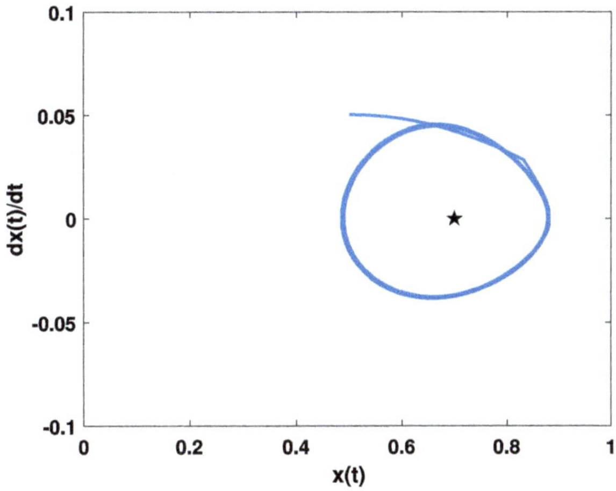

Fig. 4 Two-dimensional projection of the phase portrait for the example from Fig. 3. The phase portrait is drawn in the $x - \dot{x}$ plane

does exceed the critical Hopf time delay of $\tau_c \approx 7.48$ predicted by the analysis. The simulated dynamics for this Hawk–Dove imitation game beginning with an equal proportion of cooperators and defectors (i.e. $x(0) = 0.5$) is shown in Fig. 3, with its corresponding phase portrait drawn in Fig. 4. In this scenario, the increased time delay has led to a de-stabilization of the Nash equilibrium at x^*. Because of the increased time delay, individual players make decisions based on information that is too old to be of value, causing an over-reaction that initially creates too many cooperators. However, by the time that these cooperators realize there are too many cooperators, there are many who have already changed to defection (which they are unaware of because of the time delay), thus causing too many defectors. This cyclical over-reaction repeats itself leading to a limit-cycle behavior as seen in the phase portrait in Fig. 4. While the limit cycle behavior is an interesting form of an unstable behavior, the important feature that we are examining in this chapter is the destabilization of the equilibrium, and the limit cycle behavior is only guaranteed for values of time delay just over the critical time delay. It is an interesting analysis for future work to investigate the nature of the unstable behavior for larger values of time delays. For this analysis, it is important to note that the transition point for where the change in the asymptotic stability of the Nash equilibrium occurs was correctly predicted by the Hopf bifurcation analysis.

3 The Aspiration and Imitation Game Dynamics

While the imitation game rules presume individual players adjust their strategies in comparison with other players, there are other aspirational game rules in which players adjust strategies in an attempt to reach a stated payoff goal. In the most basic form of aspirational games, the players individually change their strategy (i.e. the individual switches from C to D or from D to C) whenever their payoff is below an aspiration threshold K [11]. This type of strategy can be represented as a degenerate form of the imitation game whereby the imitation rule $\psi(u, v)$ only depends on the payoff of an individual player (i.e. where $\psi(u, v) = \psi(u; K)$). However, this aspiration-only game process lacks the social cohesion found in most physical and social group systems. In an attempt to gain some of that social cohesion, various dynamic aspirational games have been developed where the aspiration level K varies over time in a manner that depends on the status of the overall population (i.e. $K(t) = f(x(t))$) [13, 14]. While such models have interesting dynamics and are worthy of future study, we are concerned instead with games that modify the imitation process by only adding a basic aspirational component.

An approach to developing a combination of aspiration game process within the interactions of an imitation game was developed by Gale et al. [6] and described by Samuelson [15]. The evolutionary game process that they propose as a variant of the imitation game dynamics is referred to as the *aspiration and imitation process*. In their aspiration and imitation model, players first decide at each time step whether or not to “learn” based on their recent history. A non-learning player continues their current strategy, whereas a learning player compares their recent payoff to an aspirational threshold level, $K > 0$. If they meet or exceed the threshold, they maintain their current strategy. However, if that fall short of the threshold level, they then interact with a randomly chosen player and either adopt that player’s strategy (with probability γ) or adopt the strategy opposite of that player (with probability $1 - \gamma$).

For a large population of players with all players following the rules of such a game over a long (but finite) interval of time, the resulting dynamics are given as follows [15]:

$$\frac{dx(t)}{dt} = x(t)\left[\pi_C(t) - \bar{\pi}(t)\right] + \gamma\left[1 - 2x(t)\right]\left[K - \bar{\pi}(t)\right] \tag{32}$$

Note that in the case of $\gamma = 0$, the aspiration and imitation game dynamics reduce to the following:

$$\begin{aligned}\frac{dx(t)}{dt} &= x(t)\,[\pi_C(t) - \bar{\pi}(t)] \\ &= x(t)\,[\pi_C(t) - (x(t)\,\pi_C(t) + (1 - x(t))\,\pi_D(t))] \\ &= x(t)\,(1 - x(t))\,\,[\pi_C(t) - \pi_D(t)]\end{aligned} \tag{33}$$

which is the imitation game dynamics of Eq. (12). Thus, the term with γ in Eq. (32) can be thought of as an additional dynamic that accounts for the aspirational effect in the aspiration and imitation game.

Consider now the aspiration and imitation game dynamics for the Hawk–Dove game with payoffs of the form (4). Recall that the imitation game dynamics of Eq. (16) (i.e. without aspiration) lead to an equilibrium payoff of $\bar{\pi}^* = (1 - \rho^2)$ (see Eq. (18)) for the equilibrium at $x^* = (1 - \rho)$. The portion of the aspiration and imitation game dynamics of Eq. (32) that is attributable to the aspiration behaves by drawing towards values of $x(t) \to 0.5$, while also trending the expected payoff $\bar{\pi}(t) \to K$. Therefore, choosing an aspirational value of $K = (1 - \rho^2)$ leads to an aspiration that does not increase nor decrease the natural equilibrium value of the imitation game dynamics. Values of $K > (1 - \rho^2)$ create an aspiration that pushes the equilibrium value larger than that for the native imitation game dynamics, while values of $K < (1 - \rho^2)$ create an aspiration for an equilibrium lower than that for the native imitation game dynamics. For this reason, we focus on $K = (1 - \rho^2)$ to maintain the same equilibrium value of $x^* = (1 - \rho)$.

Applying the aspiration level of $K = (1 - \rho^2)$ and using the time-delayed form for $\bar{\pi}(t)$ to account for the information delay (as given in Eq. (17)), the time-delayed form of the aspiration and imitation game dynamics for the Hawk–Dove game becomes

$$\begin{aligned}\frac{dx(t)}{dt} &= x(t)\,(1 - x(t))\,\,[(1 - \rho) - x(t - \tau)] + \gamma\,(1 - 2x(t)) \times \\ &\quad \left[(1 - \rho^2) - (1 - \rho)x(t) - (1 + \rho)x(t - \tau) + x(t)x(t - \tau)\right]\end{aligned} \tag{34}$$

We note that this model is only valid for small values of γ. As γ gets very large, the imitation portion of the ensemble dynamics becomes negligible, and the aspiration-only process does not contain the bounding unstable equilibrium points at $x = \{0, 1\}$, which are necessary to keep the modeled system dynamics in the physical realm. We thus restrict our analysis to small values of $\gamma \geq 0$.

3.1 Stability Analysis of the Aspiration and Imitation Game

In the absence of a time delay (i.e. when $\tau = 0$), the game dynamics of the Hawk–Dove form of the aspiration and imitation game (Eq. (34)) takes the form

$$\frac{dx(t)}{dt} = x(t)\,(1 - x(t))\,[(1-\rho) - x(t)] + \gamma\,(1 - 2x(t))\,\left[(1 - x(t))^2 - \rho^2\right] \tag{35}$$

The only equilibrium for Eq. (35) with $\gamma > 0$ is found at

$$x^* = (1 - \rho) \tag{36}$$

which corresponds to the mixed strategy equilibrium of the imitation game (note that the aspiration and imitation game dynamics of Eq. (35) reduces to the imitation game dynamics when $\gamma = 0$). We thus focus the remainder of our analysis on this equilibrium value.

For the equilibrium at $x^* = (1 - \rho)$, the local linearized form of the dynamics are given by forming $\xi(t) = x(t) - (1 - \rho)$, and maintaining only terms linear in $\xi(t)$ (for the small neighborhood assumption), to obtain

$$\frac{d\xi(t)}{dt} = (\rho^2 - \rho)\xi(t) + 2\gamma(\rho^2 - 2\rho)\xi(t) + O\left(|\xi(t)|^2\right) \tag{37}$$

This equilibrium is asymptotically stable for values of ρ and γ such that

$$(\rho^2 - \rho) + 2\gamma(\rho^2 - 2\rho) < 0 \tag{38}$$

which, for $\rho > 0$, simplifies to

$$\rho < 1 + \frac{2\gamma}{1 + 2\gamma} \tag{39}$$

However, it is stated that $\rho < 1$, by definition, so the expression holds for any $\gamma \geq 0$ in the domain of interest. Thus, we have that the equilibrium at $x^* = (1 - \rho)$ is asymptotically stable for all values of $\rho \in (0, 1)$ for any $\gamma \geq 0$. Recall that the case of $\gamma = 0$ reduces to the dynamics of the imitation game, and under that condition, the other two unstable equilibria (at $x = 0$ and $x = 1$) re-appear. As before, we recognize that the presence of a time delay may cause asymptotically stable equilibria such as this mixed strategy equilibrium to de-stabilize through a Hopf bifurcation [4]. In the next section, we examine when such a bifurcation may occur for this system.

3.2 *Hopf Bifurcation Due to the Time Delay*

We consider the delayed aspiration and imitation game dynamics (Eq. (34)) in a neighborhood of the mixed strategy equilibrium at $x^* = (1 - \rho)$. To form a characteristic equation for this delayed differential equation, we again form two local variables in a neighborhood of the equilibrium point, one given by $\xi(t) = x(t) - (1 - \rho)$ and the second incorporating the delay as $\xi^d(t) = \xi(t - \tau) = x(t - \tau) - (1 - \rho)$.

Replacing the terms $x(t)$ and $x(t-\tau)$ in Eq. (34) with their local counterparts, we arrive at

$$\frac{d\xi(t)}{dt} = (1-\rho+\xi(t))(\xi(t)-\rho)\xi^d(t) + \gamma\left[(2\rho-1)-2\xi(t)\right](\xi(t)-2\rho)\xi^d(t) \tag{40}$$

Keeping only terms that are linear in the local variables leads to the linearized form of the local aspiration and imitation game dynamic equation as

$$\frac{d\xi(t)}{dt} = -\rho(1-\rho)\xi^d(t) + 2\gamma\rho(1-2\rho)\xi^d(t) + O(|\xi(t)|^2) \tag{41}$$

where we have used the fact that $\xi(t)$ and $\xi^d(t)$ are of that same order of magnitude (that is, $O(|\xi(t)|) = O(|\xi^d(t)|)$ since they only differ by a constant time delay). Introducing the characteristic parameter λ through $\xi(t) = e^{\lambda t}$ and $\xi^d(t) = e^{\lambda(t-\tau)}$ creates a characteristic equation of the form

$$\begin{aligned}\lambda e^{\lambda\tau} &= -\rho(1-\rho) + 2\gamma\rho(1-2\rho)\\ &= (2\gamma-1)\rho + (1-4\gamma)\rho^2\end{aligned} \tag{42}$$

The stability properties of the equilibrium point at x^* are thus given by the location of the values of λ in the complex plane.

From the previous section, it was shown that the equilibrium at x^* is asypmtotically stable for all values of $\rho \in (0,1)$ and $\gamma \geq 0$ when $\tau = 0$. To see if and when instability due to a Hopf bifurcation occurs as the time delay τ is increased from zero, we look for a crossing of the values of λ into the right-hand side of the complex plane. Thus, we set $\lambda = i\omega$ in the characteristic equation to correspond to this transition point at a critical value of the time delay of $\tau = \tau_c$. Making that substitution and collecting real and imaginary terms leads to the following two equations:

$$\text{Real:} \quad 0 = \left[(2\gamma-1)\rho + (1-4\gamma)\rho^2\right]\cos(\omega\tau_c) \tag{43}$$

and

$$\text{Imaginary:} \quad \omega = -\left[(2\gamma-1)\rho + (1-4\gamma)\rho^2\right]\sin(\omega\tau_c) \tag{44}$$

By re-writing Eq. (43) for $\cos(\omega\tau_c)$ and Eq. (44) for $\sin(\omega\tau_c)$ and applying the Pythagorean identity, we arrive at an expression in terms of ω which can be solved as

$$\omega = \left|(2\gamma-1)\rho + (1-4\gamma)\rho^2\right| \tag{45}$$

We note that the absolute value found in Eq. (45) is not necessary in practice, since

$$\begin{aligned}(2\gamma - 1)\,\rho + (1 - 4\gamma)\,\rho^2 &= -(1 - 2\gamma)\,\rho + (1 - 2\gamma)\,\rho^2 - 2\gamma\rho^2 \\ &= \ (1 - 2\gamma)\,(\rho^2 - \rho) - 2\gamma\rho^2 \\ &< \ (1 - 2\gamma)\,(\rho^2 - \rho) \\ &< \ 0 \quad \text{for} \quad \gamma \in [0, 0.5) \quad \text{and} \quad \rho \in (0, 1) \end{aligned} \tag{46}$$

Therefore, for practical values of $\gamma \in [0, 0.5)$, we have that Eq. (45) reduces to

$$\omega = (1 - 2\gamma)\rho - (1 - 4\gamma)\rho^2 \tag{47}$$

Combining this value of ω with the fact that $\cos(\omega\tau_c) = 0$ (see Eq. (43)) leads to the critical time delay at the Hopf bifurcation of

$$\tau_c = \frac{\pi}{2(1 - 2\gamma)\rho - 2(1 - 4\gamma)\rho^2} \tag{48}$$

which is valid for all $\rho \in (0, 1)$ and $\gamma \in [0, 0.5)$. Thus, we expect that for time delays which are less than the critical value from Eq. (48), the aspiration and imitation game dynamics should lead to the asymptotically stable value of x^*. However, for larger values of $\tau > \tau_c$, we expect that the aspiration and imitation game dynamics should show the equilibrium value to be unstable. In the next section, we show numerical verification of this transition.

3.3 Numerical Example

The dynamics of the Hawk–Dove aspiration and imitation game with delays (as given in Eq. (34)) were simulated numerically to assess the system's long-term dynamic behavior. As with the imitation game, we are interested in observing when and how the fraction of cooperators $x(t)$ leads to the mixed strategy Nash equilibrium of $x^* = 1 - \rho$. Consider a Hawk–Dove aspiration and imitation game with delays and a benefit-to-cost parameter value of $\rho = 0.3$ and an aspiration parameter value of $\gamma = 0.1$. The mixed strategy Nash equilibrium corresponding to this value of ρ occurs at $x^* = 1 - \rho = 0.7$. From the analysis of the preceding section, an expected Hopf bifurcation for the asymptotic stability of this mixed strategy equilibrium should occur at $\tau_c \approx 8.45$ (see Eq. (48)). Assume a small time delay of $\tau = 8.0$ (considered small because it is below the expected critical Hopf delay τ_c). The simulated dynamics for a Hawk–Dove imitation game beginning with an equal proportion of cooperators and defectors (i.e. $x(0) = 0.5$) is shown in Fig. 5, with a corresponding phase portrait drawn in Fig. 6. Clearly the numerical solution demonstrates asymptotic stability of the proportion of cooperators approaching the Nash equilibrium.

Next consider the same Hawk–Dove aspiration and imitation game with a slightly larger time delay of $\tau = 9.0$. While this time delay is only slightly larger than the prior case, it does exceed the critical Hopf time delay of $\tau_c \approx 8.45$ predicted by the

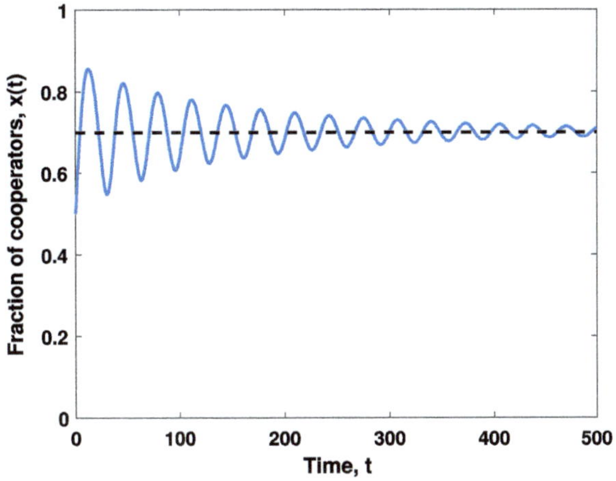

Fig. 5 Time series for the aspiration and imitation game dynamics with $\rho = 0.3$, $\gamma = 0.1$ and a time delay of $\tau = 8$. The dashed line is the Nash equilibrium solution at $x^* = 0.7$

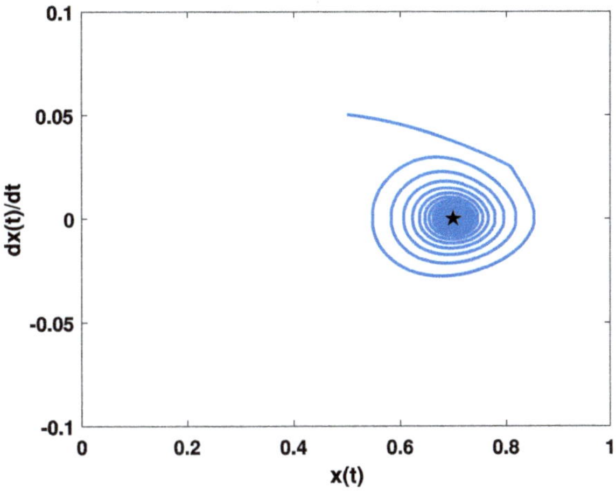

Fig. 6 Two-dimensional projection of the phase portrait for the example from Fig. 5. The phase portrait is drawn in the $x - \dot{x}$ plane

analysis. The simulated dynamics for this Hawk–Dove imitation game beginning with an equal proportion of cooperators and defectors (i.e. $x(0) = 0.5$) is shown in Fig. 7, with its corresponding phase portrait drawn in Fig. 8. Just like in the imitation game, the increased time delay has led to a de-stabilization of the Nash equilibrium at x^*, as predicted by the Hopf bifurcation analysis.

We also note that the value of γ affects how much of an effect the addition of aspirational behavior adds to the sensitivity to time delays. In particular, from Eq. (48) it can be shown that for $\rho < 0.5$, we have $d\tau_c/d\gamma > 0$, such that increasing the amount of aspirational effect will increase the critical time delay τ_c. In particular, note that

$$\frac{d\tau_c}{d\gamma} = \frac{4}{\pi}\,\tau_c^2 \rho\,(1 - 2\rho) \tag{49}$$

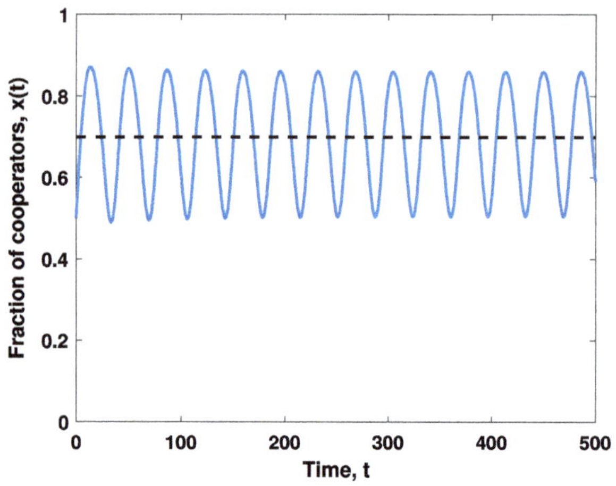

Fig. 7 Time series for the aspiration and imitation game dynamics with $\rho = 0.3$, $\gamma = 0.1$ and a time delay of $\tau = 9$. The dashed line is the Nash equilibrium solution at $x^* = 0.7$

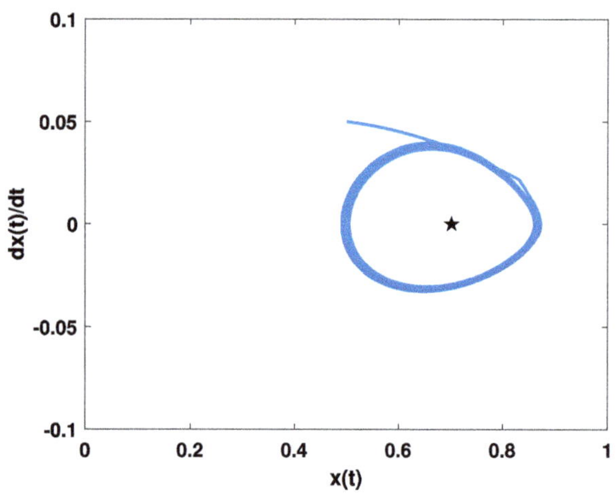

Fig. 8 Two-dimensional projection of the phase portrait for the example from Fig. 7. The phase portrait is drawn in the $x - \dot{x}$ plane

Thus, the population's evolutionary tolerance to time delays can be improved by the addition of more aspirational behavior. Note that this effect is only valid for low values of game reward where $v < 0.5c$ (which corresponds to $\rho < 0.5$). As the game reward value increases such that $v > 0.5c$ (corresponding to $\rho > 0.5$), the tolerance for longer time delays actually decreases with increasing aspirational effect. This effect is illustrated in Fig. 9, which shows the critical time delay values τ_c as a function of ρ for various amounts of aspirational effect (given by γ).

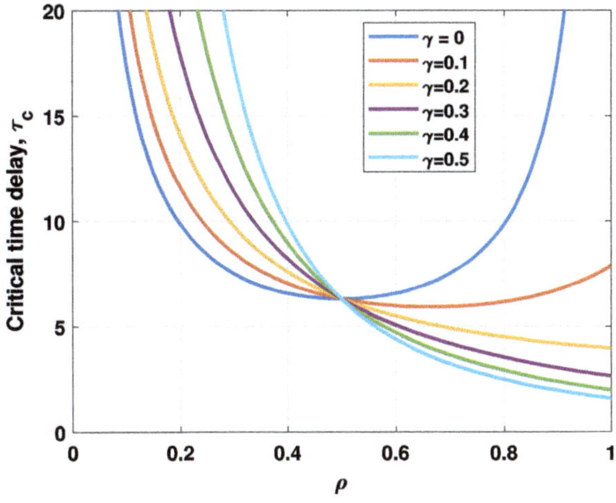

Fig. 9 Plot of the critical time delay τ_c as a function of ρ for various values of γ. Note that the $\gamma = 0.0$ curve is for the case with no aspirational behavior (i.e. imitation game dynamics only)

4 Conclusion

We have presented an analysis of the impact of time delays on the dynamics of a Hawk–Dove evolutionary game with both imitation game dynamics as well as aspiration and imitation game dynamics. In both situations, a large enough time delay was shown to lead to instability of the mixed-strategy equilibrium point, causing limit cycle behavior in the long-term dynamics. A Hopf bifurcation analysis was performed to determine the critical value of the time delay at which the transition from asymptotic stability occurs. In certain situations, it was shown that the addition of the aspirational dynamics causes the aspiration and imitation game to have a higher tolerance to time delays than the imitation-only game. Numerical results were presented to validate the analysis claims.

References

1. Alboszta, J., and J. Miekisz. 2004. Stability of evolutionarily stable strategies in discrete replicator dynamics with time delay. *Journal of Theoretical Biology* 231: 175–179.
2. Arefin, M. R., and J. Tanimoto. 2021. Imitation and aspiration dynamics bring different evolutionary outcomes in feedback-evolving games. *Proceedings of the Royal Society A* 477: 20210240.
3. Deressa, C. T., and D. T. Etefa. 2021. Parameter based stability analysis of generalized mathematical model with delay of competition between two species. *Applied Mathematics and Computation* 394: 125791.
4. Erneux, T. 2009. *Applied Delay Differential Equations*. Springer.
5. Etesami, S. R. 2019. A simple framework for stability analysis of state-dependent networks of heterogeneous agents. *SIAM Journal on Control and Optimization* 57 (3): 1757–1782.
6. Gale, J., K. G. Binmore, and L. Samuelson. 1995. Learning to be imperfect: The ultimatum game. *Games and Economic Behavior* 8: 56–90.

7. Hofbauer, J., and K. Sigmund. 1998. *Evolutionary Games and Population Dynamics*. Cambridge University Press.
8. Hofbauer, J., and K. Sigmund. 2003. Evolutionary game dynamics. *Bulletin of the American Mathematical Society* 40 (4): 479–519.
9. Mittal, S., A. Mukhopadhyay, and S. Chakraborty. 2020. Evolutionary dynamics of the delayed replicator-mutator equation: Limit cycle and cooperation. *Physical Review E* 101: 042410.
10. Obando, G., J. I. Povedo, and N. Quijano. 2016. Replicator dynamics under perturbations and time delays. *Mathematics of Control, Signals, and Systems* 28: 20.
11. Palomino, F., and F. Vega-Redondo. 1999. Convergence of aspirations and (partial) cooperation in the prisoner's dilemma. *International Journal of Game Theory* 28 (4): 465–488.
12. Platkowski, T. 2009. Enhanced cooperation in prisoner's dilemma with aspiration. *Applied Mathematics Letters* 22: 1161–1165.
13. Platkowski, T., and P. Bujnowski. 2009. Cooperation in aspiration-based N-person prisoner's dilemmas. *Physical Review E* 79: 036103.
14. Posch, M., A. Pichler, and K. Sigmund. 1999. The efficiency of adapting aspiration levels. *Proceedings of the Royal Society B* 266: 1427–1435.
15. Samuelson, L. 1997. *Evolutionary Games and Equilibrium Selection*. MIT Press.
16. Shampine, L. F., and S. Thompson. 2001. Solving DDEs in Matlab. *Applied Numerical Mathematics* 37 (4): 441–458.
17. Tao, Y., and Z. Wang. 1997. Effect of time delay and evolutionarily stable strategy. *Journal of Theoretical Biology* 187: 111–116.
18. Wang, Z., S. Kokubo, M. Jusup, and J. Tanimoto. 2015. Universal scaling for the dilemma strength in evolutionary games. *Physics of Life Reviews* 14: 1–30.
19. Weibull, J. W. 1995. *Evolutionary Game Theory*. MIT Press.
20. Wesson, E., R. Rand, and D. Rand. 2016. Hopf bifurcation in two-strategy delayed replicator dynamics. *International Journal of Bifurcation and Chaos* 26.
21. Wettergren, T. A. 2021. Replicator dynamics of an N-player snowdrift game with delayed payoffs. *Applied Mathematics and Computation* 404: 126204.

Price Formation in Financial Markets: A Mean-Field Game Perspective

David Evangelista, Yuri Saporito, and Yuri Thamsten

Abstract We propose a mean-field game model to study the price formation of an asset negotiated in an order book, considering costs stemming from limited liquidity. We derive an analytical formula for the formed price in terms of the realized order flow. Finally, we numerically assess our results with a large experiment using high-frequency data from ten stocks listed in the NASDAQ.

Keywords Price formation · Optimal trading · Mean-field games

1 Introduction

Asset price formation arises from the interactions of traders through the limit order book (LOB), [22]. Traders base their actions on the LOB's current state, which collectively determines the asset price. This perspective differs from the price discovery approach, where it is assumed that an intrinsic asset price exists due to fundamental factors, and traders' views help to reveal this true price over time, [45]. In price formation, prices are built in a forward manner, whereas price discovery involves a backward process to uncover the true price.

The literature on price formation delves into fundamental questions about market dynamics, such as the speed at which new private information affects prices, the value of private information to insiders, the impact of noise trading on price volatility, and the determinants of market liquidity, [40]. Many models have been developed to explore these issues, focusing particularly on liquidity and agent-based

D. Evangelista · Y. Saporito (✉)
Escola de Matemática Aplicada (EMAp), Fundação Getulio Vargas, Rio de Janeiro, Brazil
e-mail: yuri.saporito@fgv.br

D. Evangelista
e-mail: david.evangelista@fgv.br

Y. Thamsten
Instituto de Matemática e Estatística (IME), Universidade Federal Fluminense, Niterói, Brazil
e-mail: ythamsten@id.uff.br

E. A. Pimentel and B. Toni (eds.), *Differential and Algorithmic Intelligent Game Theory*, STEAM-H: Science, Technology, Engineering, Agriculture, Mathematics & Health,
https://doi.org/10.1007/978-3-031-97733-6_3

models, [13]. Statistical investigations of order books have also yielded significant insights, especially from researchers applying principles from physics to financial markets, a field known as econophysics, [6, 14, 41, 50].

On the other hand, the price discovery literature examines how market microstructure properties influence the process of uncovering true asset prices, [16, 25, 35–37]. Key studies have focused on econometric measures to analyze price discovery, [5, 15, 19, 30, 31, 34], and have investigated how assets traded across multiple venues integrate new information. These studies aim to identify which markets lead in the incorporation of novel information into asset prices, [34], and [18, 20, 24, 32, 46].

In the financial industry, understanding price formation is crucial, as it links the order flow to price changes, a relationship often described by Kyle's lambda, [40]. This knowledge is especially relevant in studying price impact, which examines how order flow affects asset price dynamics. Recently, research has also extended to electricity markets, where price formation is influenced by factors such as renewable energy certificates and market clearing conditions, [51]. Overall, understanding price formation provides valuable insights into market behavior and has practical applications in areas like transaction cost analysis, [8, 9, 47].

1.1 Contribution and Outline of the Paper

To address liquidity issues, we build upon the frameworks established by [1, 4], which differentiate between permanent and temporary price impacts. Permanent impacts affect the price dynamics over time, while temporary impacts result from immediate execution of orders. The concept of "order book resilience" assumes that the book replenishes instantly after a temporary impact. Subsequent advancements, such as those by [27, 48], expanded this idea to include transient price impacts. We aim to extend these models to explore price formation, incorporating temporary price impacts without assuming a specific shape for their influence on asset price dynamics. Instead, we explain asset prices based on realized order flow.

Assuming there are infinitely many players, we consider a Mean-Field Game (MFG) to investigate equilibria in large competitive populations. MFGs, introduced by Lasry and Lions [42–44], and by Huang and Caines and Malhamé [38, 39], offer analytic tractability. We introduce the notion of competitive equilibria for our problem. Through this definition, we can analyze the interplay between the supply/demand resulting from the order flow and the corresponding asset price. We assume continuity of the price and order flow, as well as that they are deterministic, working in an MFG framework akin to that of [29] (see [2] for connections with Aubry-Mather theory to understand the long term behavior of players and [28] in which the supply is a stochastic process). Although our assumptions are strong at this point, we will see that our results are satisfactory empirically. Furthermore, we weaken our assumptions in a finite player game setting in [23].

We organize the remainder of this paper as follows. In Sect. 2, we analyze competitive equilibria in our MFG setting and derive our main theoretical results. Section 3 numerically assesses our results from the previous section with high-frequency data from NASDAQ. We conclude the paper in Sect. 4.

2 Price Formation with a MFG Model

2.1 The Model

We consider a single venue in which infinitely many agents trade a single asset. By this, we mean that their trading in other assets do not influence their decisions in the specific one we are focusing on. We denote the asset price by p. Our point of view is that p is the result of the interaction among the population through the order book. As a simplifying assumption, we stipulate that p is deterministic, as in [29]. In our related article [23], we relax this restriction in the finite population setting, and we provide an appropriate interpretation for the price we find in the present work as a limit when the number of agents goes to infinity.

We fix a complete probability space $(\Omega, \mathcal{F}, \mathbb{P})$, supporting a one-dimensional Brownian motion W, and consider the filtration generated by it, $\mathbb{F} = \{\mathcal{F}_t\}_{0 \leq t \leq T}$, with $\mathcal{F} = \mathcal{F}_T$. Therefore, we currently suppose symmetric information throughout the population.

Let us now consider a representative trader beginning a trading program at time $t \in [0, T]$, targeting to execute a trade of size $|q|$ shares of the asset, where $q \in \mathbb{R}$. If $q > 0$ (resp., $q < 0$) the agent seeks to liquidate (resp., execute) q shares over $[t, T]$. The trader controls her turnover rate $\{\nu_u\}_{t \leqslant u \leqslant T} \in \mathcal{U}_t$, where

$$\mathcal{U}_t := \left\{ \nu = \{\nu_u\}_{t \leqslant u \leqslant T} : \nu \text{ is } \mathbb{F} - \text{progressively measurable, and } \mathbb{E}\left[\int_t^T \nu_u^2 \, du\right] < \infty \right\}$$

is the set of admissible controls. In this manner, her inventory process $\{q_u\}_{t \leqslant u \leqslant T}$ satisfies

$$\begin{cases} dq_u = \nu_u \, du + \sigma \, dW_u, \\ q_t = q. \end{cases} \tag{1}$$

Above, we allowed uncertainty in the traders' inventory. As put by [21], this is justified because it represents an effect observed in reality. If the trader is a broker, she will receive an order flow from her clients, something which we can model as (1), cf. [11, Sect. 2.1]. Another explanation for introducing uncertainty in the dynamics (1) of the inventory is that when the agent actually conducts her trades, at the so-called tactical layer of an execution algorithm, in the terminology of [33], she is subject to execution uncertainty stemming from operational risks. However, one should notice that we could consider $\sigma = 0$ without any impact in the arguments below. In particular, the formed price p will not be affected by the volatility σ, cf. Eq. (15).

As the trader executes a volume $\nu_u \, du$ over the time interval $[u, u + du[$ she must pay or receive—depending on the sign of ν_u— the quantity $-p(u)\nu_u \, du$. If we assume a linear temporary price impact, then the indirect cost per share she incurs from this effect is $-\kappa \nu_u$; hence, this trade's total cost is $-\kappa \nu_u^2 \, du$. Gathering these two pieces together, we derive the following dynamics for the corresponding cash process $\{c_t\}_{0 \leqslant t \leqslant T}$:

$$\begin{cases} dc_u = -p(u)\nu_u \, du - \kappa \nu_u^2 \, du = -\left[p(u) + \kappa \nu_u\right] \nu_u \, du, \\ c_t = c. \end{cases} \tag{2}$$

In (2), we denoted by $c \in \mathbb{R}$ the initial cash amount that the trader holds. We remark that we assume κ to be the same among all players. Since we suppose they are all trading the same asset in the same venue, this is not an unrealistic assumption.

Finally, we define the agent's wealth w as

$$w_u := c_u + q_u p(u) = c - \int_t^u \left[p(\tau) + \kappa \nu_\tau\right] \nu_\tau \, d\tau + q_u p(u) \qquad (t \leqslant u \leqslant T). \tag{3}$$

We intend to approach the price formation in the MFG model via the dynamic programming perspective. To conduct this endeavor, we define the value function V as

$$\begin{aligned} V(t, q) &:= \sup_{\nu \in \mathcal{U}_t} \mathbb{E}\left[w_T - c - A\, q_T^2 - \frac{\phi}{2} \int_t^T q_u^2 \, du \,\middle|\, c_t = c,\, q_t = q \right] \\ &= \sup_{\nu \in \mathcal{U}_t} \mathbb{E}\left[q_T \left(p(T) - A q_T\right) - \int_t^T \left\{ \kappa\, \nu_u^2 + \frac{\phi}{2} q_u^2 + \nu_u p(u) \right\} du \,\middle|\, q_t = q \right]. \end{aligned} \tag{4}$$

Here, V is the supremum of performance criteria which are in turn formed by three parts. Firstly, they include the mean of the difference $w_T - c$ between the corresponding trader's terminal wealth and initial cash. We do not compare $w_T - c$ to the pre-trade price $q\, p(t)$ of the initial inventory, a common enterprise in *Implementation Shortfall* algorithms, because we are not interested in taking differentials of the price. Secondly, the criteria also include a terminal penalization for finishing with non-vanishing inventory, i.e., $-A\, q_T^2$. Thirdly, there is the mean of an urgency term $-\phi \int_t^T q_u^2 \, du$, inducing traders to accelerate their program. Analogously, as authors in [12] explain, we emphasize that we do not interpret this as a financially meaningful penalization but rather as an extra risk management tool. However, as put by [21] in a related modeling context, even if we assume traders are risk-neutral, they can incur costs for holding non-zero inventory, either long or short, which advocates in part of taking ϕ to be the same for all agents.

The final ingredient of the mean-field game model is the density of the law of the players' inventories at any given time t we denoted by $m(t, \cdot)$. Additionally, we consider the initial density m_0 as given, i.e., $m(0, \cdot) = m_0$.

If an agent with inventory level q trades at a rate $\nu(t, q)$, then we can define the aggregation rate μ as

$$\mu(t) = \int_{\mathbb{R}} \nu(t, q) m(t, q)\, dq.$$

To clear the market, we consider that a market demand/supply rate $\Lambda : [0, T] \to \mathbb{R}$ is given and the following condition is satisfied, cf. [29]:

$$\mu(t) + \Lambda(t) = 0 \qquad (0 \leqslant t \leqslant T). \tag{5}$$

Therefore, if the market is bullish (respectively, bearish) on aggregate, then we would expect $\mu(t) > 0$ (respectively, $\mu(t) < 0$), i.e., traders are required to buy (respectively, sell) some shares of the asset; hence, $\Lambda(t)$ describes the opposite movement, which occurs in the inventory of the traders' counterparts as a whole. We thus see Λ as the interaction process of agents with the LOB itself (such as by adding or subtracting shares of an asset from there), or rather the resulting movement in the LOB due to the negotiations taking place there. We remark that we do not suppose that the players behaving optimally in our MFG trade among themselves—in fact, players trading aggressively (meaning via market orders) could be, say, all buying the asset simultaneously at a certain time instant. Thus, we do implicitly regard that there are exogenous liquidity providers accepting to serve as counterparts at price p to the mean-field of traders we are studying.

2.2 *Competitive Equilibria*

We are now ready to provide the following definition of competitive equilibria for the MFG model for a given market demand/supply rate $\Lambda : [0, T] \to \mathbb{R}$.

Definition 2.1 A competitive equilibrium for the MFG model is a tuple (p, V, m, ν^*), where $p : [0, T] \to \mathbb{R}$ and $V, m, \nu^* : [0, T] \times \mathbb{R} \to \mathbb{R}$ satisfy:

(a) $p \in C[0, T]$;
(b) For each square-integrable random variable ξ, the SDE $dq_u^* = \nu^*(u, q_u^*)\, du + \sigma\, dW_u$, $q_t^* = \xi$, has a strong solution;
(c) For each $(t, q) \in [0, T] \times \mathbb{R}$, if we consider a solution of the SDE above with initial condition $q_t^* = q$ and write $\nu^* := \left\{\nu_u^* := \nu^*(u, q_u^*)\right\}_{t \leqslant u \leqslant T}$, then the supremum in (4) is attained by the strategy ν^*, and it is equal to $V(t, q)$;
(d) If the density of Law(q_0^*) is m_0 and $\left\{q_t^*\right\}_{0 \leqslant t \leqslant T}$ is a solution of SDE in (a) with such initial condition, then the function m is the inventory distribution density of q^*, i.e. $m(t, \cdot)$ is the density of Law(q_t^*), for $0 \leqslant t \leqslant T$;
(e) Equation (5) holds for ν^* and m:

$$\int_{\mathbb{R}} \nu^*(t, q) m(t, q)\, dq + \Lambda(t) = 0 \qquad (0 \leqslant t \leqslant T). \tag{6}$$

Condition (c) is akin to the "*Profit Maximization*" property of the equilibria [40]. On the other hand, we do not propose "market efficiency" in the context of our equilibria configurations. Rather, we leave a degree of freedom, viz., the order flow of liquidity taking orders. In our condition (e), we assume market makers clear the market, but following a perspective similar to that of [40],[1] i.e., we do not delve into the (interesting) question of their optimizing behavior in doing so.

2.3 Some Properties of the Competitive Equilibria

We now turn to the analysis of competitive equilibria in terms of Definition 2.1. The first step is to derive the well-posedness for (V, m). Particularly, the linear-quadratic shape of the criteria in (4) will allow us to show that V is itself quadratic in the inventory variable. These comprise the content of the subsequent result.

Proposition 2.2 *Let us consider a competitive equilibrium* (p, V, m, ν^*). *Then,* (V, m) *must solve the system*

$$\begin{cases} \partial_t V(t,q) + \frac{\sigma^2}{2}\partial_q^2 V(t,q) + \frac{1}{4\kappa}\left[\partial_q V(t,q) - p(t)\right]^2 - \frac{\phi}{2}q^2 = 0, \\ \partial_t m(t,q) - \frac{\sigma^2}{2}\partial_q^2 m(t,q) + \partial_q\left[\frac{(\partial_q V(t,q) - p(t))}{2\kappa} m(t,q)\right] = 0, \\ V(T,q) = q\left[p(T) - Aq\right] \text{ and } m(0,q) = m_0(q), \end{cases} \tag{7}$$

where we understand the first PDE above in the viscosity sense, and the second one in the distributional sense. Moreover, for such p, *the system (7) admits a unique smooth solution* $(V, m) \in \left[C^{1,2}\right]^2$, *where*

$$V(t,q) = \theta_0(t) + \theta_1(t)q + \theta_2(t)q^2, \tag{8}$$

with

$$\theta_2(t) = \sqrt{\kappa\phi}\frac{1 - ce^{2\gamma(T-t)}}{1 + ce^{2\gamma(T-t)}}, \text{ for } c := \frac{\sqrt{\kappa\phi/2} + A}{\sqrt{\kappa\phi/2} - A} \text{ and } \gamma := \sqrt{\frac{\phi}{2\kappa}}, \tag{9}$$

$$\theta_1(t) = p(T)e^{\frac{1}{\kappa}\int_t^T \theta_2(s)ds} - \frac{1}{\kappa}\int_t^T \theta_2(s)p(s)e^{\frac{1}{\kappa}\int_t^s \theta_2(\tau)d\tau}ds, \tag{10}$$

[1] "Modelling how market makers can earn the positive frictional profits necessary to attract them into the business of market making is an interesting topic which takes us away from our main objective of studying how price formation is influenced by the optimizing behavior of an insider in a somewhat idealized setting.", see [40, Sect. 2].

and

$$\theta_0(t) = \int_t^T \left[\frac{1}{4\kappa} (\theta_1(u) - p(u))^2 + 2\sigma^2\theta_2(u) \right] du. \tag{11}$$

Proof Since $p \in C([0, T])$, the value function V defined in (4) solves in the viscosity sense the Hamilton-Jacobi-Bellman equation comprising the first PDE in (7), see [49, Theorems 4.3.1 and 4.3.2, Remarks 4.3.4 and 4.3.5]—the terminal condition is clear from (4). Now, we find (arguing as in [10, Lemma 3.3]) that the density of players' inventory m (weakly) solves the Kolmogorov-Fokker-Planck equation in (7), with initial data $m(0, \cdot) = \text{Law}(q_0)$. To prove the second part of the proposition, we make the ansatz

$$V(t, q) = \theta_0(t) + \theta_1(t)q + \theta_2(t)q^2. \tag{12}$$

Thus, the first equation in (7) becomes

$$\begin{cases} \theta_2'(t) + \frac{1}{\kappa}\theta_2^2(t) - \phi/2 = 0, \\ \theta_1'(t) + \frac{1}{\kappa}\theta_2(t)(\theta_1(t) - p(t)) = 0, \\ \theta_0'(t) + \frac{1}{4\kappa}(\theta_1(t) - p(t))^2 + 2\sigma^2\theta_2(t) = 0, \\ \theta_0(T) = 0, \theta_1(T) = p(T), \theta_2(T) = -A. \end{cases} \tag{13}$$

Solving the Ricatti equation determining θ_2 is straightforward; it also appears in the standard Almgren-Chriss model, with only one trader, see [12]. With θ_2 at hand, we derive the formula for θ_1 by solving the second equation in (13), which is a linear ODE. Finally, we obtain θ_0 from the third one via direct integration. At this point, we can carry out standard verification arguments to validate (12). With such $V \in C^{1,2}$ at hand, showing that the Fokker-Plack-Kolmogorov equation admits a unique smooth solution m is standard, see [10]. □

Corollary 2.3 *If (p, V, m, v^*) is a competitive equilibrium, then the optimal control in feedback form is given by*

$$v^*(t, q) = \frac{\partial_q V(t, q) - p(t)}{2\kappa} = \frac{1}{2\kappa}(\theta_1(t) - p(t)) + \frac{\theta_2(t)}{\kappa}q. \tag{14}$$

The identity (14) is somewhat insightful, as it sheds light on how we can interpret the functions θ_1 and θ_2. In fact, on the one hand, since $\theta_2 < 0$, a consequence of this equation is that $\theta_2 q$ is a contribution to the trading rate of a player holding inventory q that represents a force driving her to clear these holdings. On the other hand, there is another term: $\theta_1 - p$. If $\theta_1 > p$ (respectively, $\theta_1 < p$) then this term leads the trader to buy (respectively, sell) shares. Therefore, we can interpret θ_1 as some indication of the future value of the asset—something that relation (10) itself suggests. The precise impact of the asset's future value on θ_1 is given by its definition in (10), as it depends on the values of p at times ranging from t to the final time T.

2.4 On the Formed Price in the MFG

We are now ready to state and prove the main theoretical result of our paper. Summing up, its content is a representation of the price in terms of the market supply/demand rate path Λ in a competitive equilibrium, see (15) below. Firstly, we provide the formula as a necessary condition for a given tuple to be a competitive equilibrium. In effect, this relation allows us to jump out of the model, looking for its validation on data stemming from real world prices of assets, which will be done in Sect. 3.

Theorem 2.4 *Given a competitive equilibrium* (p, V, m, ν^*)*, the formed price is*

$$p(t) = p(0) + \phi E_0 t - \phi \int_0^t (t-u)\Lambda(u)du + 2\kappa(\Lambda(t) - \Lambda(0)). \tag{15}$$

Proof Remember that the law of q_t^* is given by $m(t,\cdot)$. We define:

$$\begin{cases} \Pi(t) := \mathbb{E}[\partial_q V(t, q_t^*)] = \int_{\mathbb{R}} \partial_q V(t,q) m(t,q)\, dq; \\ E(t) := \mathbb{E}[q_t^*] = \int_{\mathbb{R}} q m(t,q)\, dq. \end{cases} \tag{16}$$

Differentiating the first equation in (7) with respect to q yields

$$\partial_q \partial_t V(t,q) + \frac{\sigma^2}{2}\partial_q^3 V(t,q) + \frac{(\partial_q V(t,q) - p(t))}{2\kappa}\partial_q V(t,q) - \phi q = 0. \tag{17}$$

We emphasize that this differentiation is licit since V is a quadratic polynomial with respect to q, with coefficients that are differentiable in time. Therefore, we observe that

$$\begin{aligned} \Pi'(t) &= \int_{\mathbb{R}} \partial_q \partial_t V(t,q) m(t,q)\, dq + \int_{\mathbb{R}} \partial_q V(t,q) \partial_t m(t,q)\, dq \\ &= \int_{\mathbb{R}} \Big\{ -\frac{\sigma^2}{2}\partial_q^3 V(t,q) - \frac{(\partial_q V(t,q) - p(t))}{2\kappa}\partial_q^2 V(t,q) + \phi q \Big\} m(t,q)\, dq \\ &- \int_{\mathbb{R}} \frac{1}{2\kappa}\partial_q V(t,q) \partial_q \left[(\partial_q V(t,q) - p(t)) m(t,q) \right] dq + \frac{\sigma^2}{2}\int_{\mathbb{R}} \partial_q V(t,q) \partial_q^2 m(t,q)\, dq. \end{aligned} \tag{18}$$

Next, we notice that carrying out integration by parts in identity (18), we find

$$\Pi'(t) = \phi E(t). \tag{19}$$

Employing (12) and (14) in (1), we infer

$$q_t^* = q_0 + \int_0^t \frac{\partial_q V(s, q_s^*) - p(s)}{2\kappa} ds + \sigma W_t,$$

in such a way that

$$E(t) = \mathbb{E}[q_0] + \int_0^t \mathbb{E}\left[\frac{\partial_q V(s, q_s^*) - p(s)}{2\kappa}\right] ds. \tag{20}$$

From (20) and (6), we deduce

$$E'(t) = \mathbb{E}\left[\frac{\partial_q V(t, q_t^*) - p(t)}{2\kappa}\right] = -\Lambda(t), \tag{21}$$

and additionally, by the definition of Π, we find

$$E'(t) = \mathbb{E}\left[\frac{\partial_q V(t, q_t^*) - p(t)}{2\kappa}\right] = \frac{\Pi(t) - p(t)}{2\kappa}. \tag{22}$$

Collecting (19), (21) and (22), we obtain

$$\begin{cases} \Pi'(t) & = \phi E(t), \\ E'(t) & = \dfrac{\Pi(t) - p(t)}{2\kappa} = -\Lambda(t). \end{cases} \tag{23}$$

On the one hand, from the second equation in system (23), we have

$$E(t) = E_0 - \int_0^t \Lambda(s) ds.$$

On the other hand, the first one gives us

$$\begin{aligned} \Pi(t) &= \Pi(0) + \phi \int_0^t E(s) ds \\ &= \Pi(0) + \phi E_0 t - \phi \int_0^t \int_0^s \Lambda(u)\, du\, ds \\ &= \Pi(0) + \phi E_0 t - \phi \int_0^t \int_u^t \Lambda(u)\, ds\, du \\ &= \Pi(0) + \phi E_0 t - \phi \int_0^t (t - u)\Lambda(u)\, du, \end{aligned}$$

as well as, again from the second equation,

$$\begin{aligned} p(t) &= \Pi(t) + 2\kappa\Lambda(t) \\ &= \Pi(0) + \phi E_0 t - \phi \int_0^t (t - u)\Lambda(u) du + 2\kappa\Lambda(t). \end{aligned} \tag{24}$$

In particular,

$$\Pi(0) = p(0) - 2\kappa\Lambda(0). \tag{25}$$

In conclusion, (24) and (25) allow us to infer

$$p(t) = p(0) + \phi E_0 t - \phi \int_0^t (t-u)\Lambda(u)du + 2\kappa(\Lambda(t) - \Lambda(0)). \tag{26}$$

□

2.5 Discussion on the Formed Price

Let us discuss the main aspects of the formula (15).

We first consider the term $2\kappa(\Lambda(t) - \Lambda(0))$—it corresponds to the contribution of the current instantaneous market supply/demand rate $\Lambda(t)$ to the formed price (at time t). We remark that it is proportional to the difference $\Lambda(t) - \Lambda(0)$, the slope being 2κ. This is the component of the price which results from the fact that traders are willing to pay costs for trading the asset. In effect, for trading a volume of $|\Lambda(t) - \Lambda(0)|$, they will pay a total of $\kappa\,|\Lambda(t) - \Lambda(0)|^2$ in transaction costs due to this market friction. The resulting influence of this movement on the formed price is $2\kappa\,(\Lambda(t) - \Lambda(0))$.

A similar consideration is valid for $2\phi E_0 t$. Namely, if the average initial holdings (or, equivalently, the initial targets) of players, E_0, is positive (resp., negative), then our modeling implies they are willing to sell (resp., buy). However, this term $2\phi E_0 t$ indicates a positive (resp., negative) drift on the price, as long as $\phi > 0$, which grows proportionally to time. Thus, if players are not quick enough to execute their schedules, we would expect prices to be moving favorably to their current holdings, with this effect being more pronounced for larger times. In a sense, this term represents an inertia component on the asset's price. Here, the inertia we refer to is distinct from what was studied in [3] (see also references therein). In our setting, this terminology only indicates that $2\phi E_0 t$ exert a sort of resistance on trading rates moving the price.

Finally, we comment on $-2\phi \int_0^t (t-u)\Lambda(u)du$, which is undoubtedly the most interesting term. We notice we can naturally interpret it as a *memory* term. Indeed, the weight $(t-u)$ of the value $\Lambda(u)$ appearing within the integral is larger for older market supply/demand rate. In a certain sense, as markets "digest" past information, they will appear more significantly in the formed price formula.

Remark 2.5 (*Dynamics of the price*) The expression for the formed price (26) implies the following dynamics for p:

$$dp(t) = \phi\left(E_0 - \int_0^t \Lambda(u)du\right)dt + 2\kappa d\Lambda(t), \tag{27}$$

where we have used the double integral representation as in the proof above. Notice the drift of p depends on the path of Λ through its integral. Moreover, one might use this dynamics to conclude the following correspondence of quadratic variations: $\langle p \rangle_t = 4\kappa^2 \langle \Lambda \rangle_t$.

Remark 2.6 (*Connection to the Transient Impact Model*) The formed price p can be rewritten as

$$p(t) = p(0) + (\phi E_0 t - 2\kappa \Lambda(0)) + \int_0^t (\phi (t-u) + 2\kappa \delta(t-u))\, dE(u),$$

where δ is the Dirac delta at 0 and we have used the fact $E'(t) = -\Lambda(t) \Leftrightarrow dE(t) = -\Lambda(t)dt$. This representation relates our model to the Transient Impact Model as presented in [7] and the continuous-time version of [26, 27]. In terms of the notation of the aforementioned literature, we find $S_t^0 = p(0) + (\phi E_0 t - 2\kappa \Lambda(0))$ and kernel $G(\tau) = (\phi \tau + 2\kappa\, \delta(\tau))$.

2.6 *Existence of an Equilibrium*

The next result show that, given the order flow Λ, we can assert that p, determined by the formula (15), is *the* price formed by this order flow.

Proposition 2.7 *Given Λ, let us define p as in (15), (V, m) as the corresponding solution of (7), and ν^* as in (14). Then (p, V, m, ν^*) is a competitive equilibrium.*

Proof Under the assumptions above, to show that (p, V, m, ν^*) is a competitive equilibrium, we need only to verify the market clearing condition (6). To do this, we integrate (14) against m, whence finding

$$\begin{aligned}\int_{\mathbb{R}} \nu^*(t,q)m(t,q)dq &= \frac{1}{2\kappa}(\theta_1(t) - p(t)) + \frac{\theta_2(t)}{\kappa}\int_{\mathbb{R}} qm(t,q)dq \\ &= \frac{\theta_1(t)}{2\kappa} + \frac{\theta_2(t)}{\kappa}E(t) - \frac{1}{2\kappa}p(t) = \frac{1}{2\kappa}(\Pi(t) - p(t)) = -\Lambda(t)\end{aligned}$$

where we used the formula for Π in (16) and its relation to Λ. □

3 Real Data Assessments

In this section we assess the performance of the formed price formula under our MFG setting using real high-frequency data. We first specify the data and the order flow measures we consider. Next, we write down the regression of price against explanatory variables that arise from the theory we developed in the previous section.

Table 1 Average and standard deviation of midprice, spread and average daily volume (ADV). The data we use in this work is from November 2014 for ten NASDAQ listed stocks. It contains all trades and snapshots of the LOB for every 100ms of the trading day

Stocks	Midprice	Spread	ADV
MU	33.4	0.01	3,325,174
	(0.823)	(0.001)	(972,993)
FB	74.68	0.01	5,616,503
	(0.893)	(0.002)	(1,381,900)
PCAR	66.57	0.02	373,722
	(0.64)	(0.009)	(122,522)
AMZN	317.18	0.11	706,172
	(13.903)	(0.049)	(214,172)
EBAY	54.11	0.01	1,224,661
	(0.664)	(0.002)	(262,809)
SMH	52.52	0.01	263,909
	(1.011)	(0.005)	(95,574)
INTC	34.5	0.01	4,717,047
	(1.104)	(0.001)	(2,069,603)
VOD	34.57	0.01	850,241
	(1.196)	(0.002)	(588,596)
GOOG	542.87	0.17	291,492
	(6.036)	(0.08)	(73,852)
MSFT	48.4	0.01	5,190,672
	(0.704)	(0.001)	(971,123)

3.1 The Data

The data we use in this work can be found at http://sebastian.statistics.utoronto.ca/books/algo-and-hf-trading/data/. It is from November 2014 for the following tickers: AMZN, EBAY, FB, GOOG, INTC, MSFT, MU, PCAR, SMH, VOD. This data contains all trades and snapshots of the LOB for every 100ms of the trading day. Table 1 displays the mean and standard deviation in parenthesis for each stock, of the midprice, spread, and average daily volume (ADV).

3.2 Order Flow Measures

From the market clearing condition (6), we see that Λ should be taken as the additive opposite of a measure of order flow. Firstly, we consider the Trade Imbalance (TI), see [17], as such a measure. We divide the trading day $[0, T]$ in 12 corresponding intervals of 30 minutes $[0 = T_0, T_1], \ldots, [T_{11}, T_{12} = 360]$, further dividing such slice into smaller intervals $T_i = t_{i,0} < \cdots < t_{i,180} = T_{i+1}$ of constant length $\Delta t = t_{i,k+1} - t_{i,k} = 10$ seconds. We then define the time series $\mathrm{TI}^i = \left\{\mathrm{TI}_k^i\right\}_{k=0}^{180}$ on $\left[T_i, T_{i+1}\right]$ as

$$\mathrm{TI}_k^i = \sum_{n=N_{i,k}+1}^{N_{i,k+1}} \left\{\mu_n^+ - \mu_n^-\right\}, \tag{28}$$

where $N_{i,k}$ denotes the number of events that have occurred on $\left[T_i, t_{i,k}\right]$, and μ_n^+ (respectively, μ_n^-) is the volume (possibly zero) of buying (respectively, selling) market orders that are traded within $\left]t_{i,n-1}, t_{i,n}\right]$, for each $n \geqslant 1$.

In order to take the LOB queue sizes into account, we also utilize a metric called Order Flow Imbalance (OFI), introduced in [17, Sect. 2.1]. The goal in using this is to consider information from the LOB first queues, since it comprises more information. We proceed to provide the definition of OFI, referring to their paper for an in-depth discussion on its design. Let us represent by $\mathfrak{p}_n^b$ and q_n^b (respectively, $\mathfrak{p}_n^a$ and q_n^a) the bid (respectively, ask) price and volume (at-the-touch), respectively, resulting from the $n-$th set of events (i.e., those occurring within the window $\left]t_{i,n-1}, t_{i,n}\right]$). We form the quantity

$$e_n := I_{\{\mathfrak{p}_n^b \geqslant \mathfrak{p}_{n-1}^b\}} q_n^b - I_{\{\mathfrak{p}_n^b \leqslant \mathfrak{p}_{n-1}^b\}} q_{n-1}^b - I_{\{\mathfrak{p}_n^a \leqslant \mathfrak{p}_{n-1}^a\}} q_n^a + I_{\{\mathfrak{p}_n^a \geqslant \mathfrak{p}_{n-1}^a\}} q_{n-1}^a, \tag{29}$$

and we define

$$\mathrm{OFI}_k := \sum_{n=N_{i,k}+1}^{N_{i,k+1}} e_n. \tag{30}$$

3.3 Regression

Considering the formed price under the MFG setting (15), we undertake the subsequent regression over each time window $\left[T_i, T_{i+1}\right]$:

$$\mathfrak{p}_k^i = a_{i,0} + a_{i,1} S_k^{i,1} + a_{i,2} S_k^{i,2} + a_{i,3} S_k^{i,3} + \epsilon_k^i, \tag{31}$$

where we ϵ_k^i is the error term, $\mathfrak{p}_k^i$ is the observed midprice, both at time $t_{i,k}$, and the time series covariates:

$$\begin{cases} S^{i,1} := \left\{S_k^{i,1} := t_{i,k}\right\}_{k=1}^{180}, \\ S^{i,2} := \left\{S_k^{i,2} := \int_0^{t_{i,k}} (t_{i,k} - u)\Lambda(u)du\right\}_{k=1}^{180}, \\ S^{i,3} := \left\{S_k^{i,3} := \Lambda(t_{i,k})\right\}_{k=1}^{180}, \end{cases} \tag{32}$$

We use the trapezoidal rule to approximate the integral in $S^{i,2}$. Moreover, we assume the regression coefficients are constant in each time window.

Although we are regressing non-stationary variables, we argue that this does not classify it as a spurious regression because of the theoretical framework behind the regression. Moreover, we also test the stationarity of the error term in the real data experiment below.

3.4 Results

We first consider that the order flow measure Λ is given by the trade imbalance with $\Lambda(t_{i,k}) = -\mathrm{TI}_k^i/\Delta t$. As a benchmark, we take a model inspired by [40], viz.,

$$\Delta \mathfrak{p}_k^i = b_{i,0} + b_{i,1}\mathrm{TI}_k^i + \widetilde{\epsilon}_k^i. \tag{33}$$

Thus, in this benchmark, we simply assume that a price movement must correspond linearly to the market trade imbalance.

In Fig. 1, we provide an example of the performance of our MFG model with parameters estimated via regression (32), together with the benchmark model (33). There, we have, in green, the formed price of the MFG using the TI order flow. The average adjusted R^2 on this trading day is given in the first column of Table 2 (cf. column R^2_{TI}). We also present the average adjusted R^2 over the whole period for all the stocks considered in this experiment in Table 3. We notice that the MFG model beats the benchmark model (cf. column $R^2_{TI_{bench}}$) systematically over all stocks we

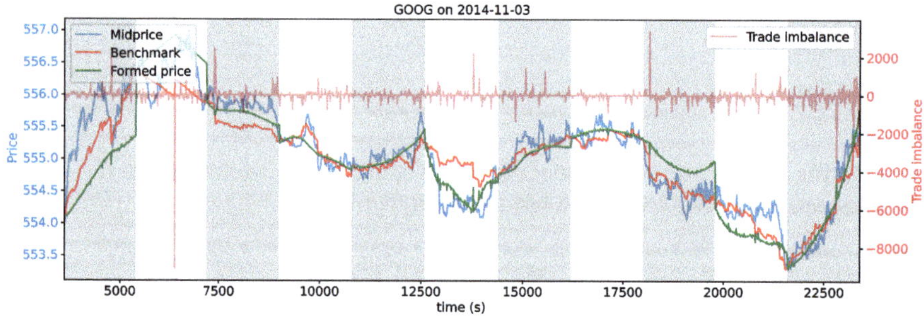

Fig. 1 Midprice and Trade Imbalance (TI) of Google (GOOG) on November 3rd, 2014, together with the estimated formed price under MFG framework and the benchmark as in Eq. (33)

Table 2 Average R^2 results of GOOG stock on November 3rd, 2014

Stock	R^2_{TI}	R^2_{OFI}	$R^2_{TI_{bench}}$	$R^2_{OFI_{bench}}$
0.9 GOOG	0.505631	0.555541	0.17395	0.459976

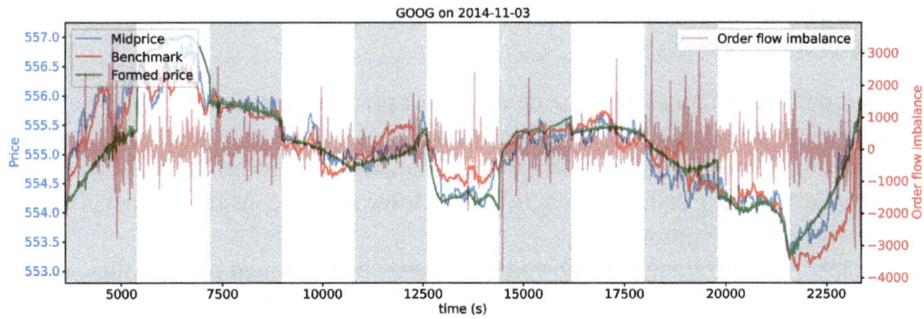

Fig. 2 Midprice and Order Flow Imbalance (OFI) of Google (GOOG) on November 3rd, 2014, together with the estimated formed price under MFG framework and the benchmark as in Eq. (33)

considered. We also point out that the adjusted R^2 achieved by the benchmark is comparable to the one observed in the literature; see [17], for instance.

To verify the statistical significance of our regression study, we provide in Table 4 the percentage of windows in which we reject the corresponding hypotheses of a given parameter being null at the 5% confidence level for the TI metric.

We now consider the order flow measure Λ given by the Order Flow Imbalance with $\Lambda(t_{i,k}) = -\mathrm{OFI}^i_k/\Delta t$. The benchmark we consider to compare our results for this metric is then

$$\Delta \mathfrak{p}^i_k = c_{i,0} + c_{i,1}\mathrm{OFI}^i_k + \overline{\epsilon}^i_k, \tag{34}$$

where the regression is performed over the slice $\left[T_i, T_{i+1}\right]$, which was studied in [17].

The performance for this order flow measure is exemplified in Fig. 2. We show in Table 5 the percentage of windows in which we can reject the hypothesis that each parameter in the regression (31) are null. We see that the use of the OFI metric leads to more statistical significance for the fourth explanatory variable (coefficient a_3), at the expense of generally weakening the second and third ones (with respective coefficients a_1 and a_2) by a bit. In Table 6 we also have the co-integration evaluation using the OFI metric instead of the TI one. We see that the TI is a better metric for the MFG model (31).

These plots indicate that using the OFI metric leads to overall better results than the trade imbalance—which would be in accordance with the results of [17]. In effect, from the results of Table 3, we notice that using the OFI metric leads to substantial improvement of the performance of the current model for this illustrative example. Moreover, model (31) outperform the current benchmark regarding the average adjusted R^2 metric. This highlights the importance of the memory term in the price formation process.

4 Conclusions

We have proposed a mean-field game model for price formation assuming informed traders had objective functionals built such that they sought to maximize their P&L, they did not want to terminate a given time horizon holding any position and an appropriate market clearing condition was in force. Moreover, we considered a penalization on non-vanishing inventory.

The theoretical framework allowed us to derived the formed price in this setting, given the market demand/supply rate, which impacts the formed price in a path-dependent way.

We provided empirical assessments for the proposed model. In them, we considered two types of order flow metrics, which also highlight the importance of such a choice in the impact of our results.

An extension of this research is conducted in [23], in which we generalize the setting presented in this paper to the finite player case. This allows us to consider stochastic prices and order flow rates. Moreover, we also establish a formed price formula that converges to the MFG formula (15) under reasonable conditions. This formula presents a new term that shows to be relevant in a similar regression analysis.

Acknowledgements YS was supported by FAPERJ (Brasil) through the Jovem Cientista do Nosso Estado Program (E-26/201.375/2022 (272760)) and by CNPq (Brasil) through the Productivity in Research Scholarship (306695/2021-9). YT was financed in part by the Coordenação de Aperfeiçoamento de Pessoal de Nível Superior—Brasil (CAPES)—Finance Code 001.

A. Tables

Table 3 Average R^2 over the whole period analyzed under each model (including benchmarks), each measure of order flow and each stock studied in the paper. We display the standard deviation of the computed R^2 in parenthesis

Symbol	MFG model		Benchmarks	
	R^2_{TI} (%)	R^2_{OFI} (%)	$R^2_{TI_{bench}}$ (%)	$R^2_{OFI_{bench}}$ (%)
AMZN	57.69	56.86	30.06	49.48
	(25.33)	(26.49)	(11.00)	(10.30)
EBAY	52.77	50.81	43.05	62.70
	(26.06)	(27.61)	(15.02)	(12.14)
FB	57.58	54.58	35.82	64.10
	(25.71)	(26.44)	(12.25)	(13.22)
GOOG	60.75	59.50	32.79	49.45
	(25.29)	(26.15)	(12.4)	(10.02)
INTC	56.51	52.67	41.44	63.04
	(25.27)	(27.22)	(11.66)	(10.68)
MSFT	57.81	52.45	44.80	69.07
	(24.42)	(27.78)	(13.3)	(10.20%)
MU	55.27	50.52	39.17	63.21
	(25.24)	(27.82)	(12.03)	(11.87)
PCAR	53.15	51.49	39.67	59.81
	(26.72)	(27.80)	(11.98)	(10.03)
SMH	56.36	52.00	12.92	54.21
	(24.29)	(24.93)	(8.92)	(11.58)
VOD	59.69	55.69	35.74	59.33
	(26.68)	(27.84)	(14.38)	(13.30)

Table 4 MFG TI hypothesis testing

Symbol	$\{a_0 \neq 0\}$ (%)	$\{a_1 \neq 0\}$ (%)	$\{a_2 \neq 0\}$ (%)	$\{a_3 \neq 0\}$ (%)
AMZN	100.00	90.28	83.80	24.07
EBAY	100.00	84.72	81.02	38.43
FB	100.00	88.89	81.48	66.20
GOOG	100.00	86.11	81.94	25.46
INTC	100.00	85.19	79.63	45.83
MSFT	100.00	89.81	85.65	43.06
MU	100.00	82.41	82.87	49.07
PCAR	100.00	86.11	75.46	31.94
SMH	100.00	84.72	84.26	18.98
VOD	100.00	86.11	77.31	25.93

Table 5 MFG OFI hypothesis testing

Symbol	$\{a_0 \neq 0\}$ (%)	$\{a_1 \neq 0\}$ (%)	$\{a_2 \neq 0\}$ (%)	$\{a_3 \neq 0\}$ (%)
AMZN	100.00	88.43	77.78	42.13
EBAY	100.00	80.09	78.24	43.52
FB	100.00	83.33	75.00	68.06
GOOG	100.00	87.04	81.48	43.06
INTC	100.00	84.26	75.93	46.30
MSFT	100.00	83.80	75.46	50.93
MU	100.00	85.19	80.09	53.70
PCAR	100.00	85.65	75.46	57.87
SMH	100.00	81.02	80.09	51.85
VOD	100.00	85.65	75.46	37.96

Table 6 Percentage of windows passing the ADF test for the 5% critical level

Symbol	MFG model	
	TI (%)	OFI (%)
AMZN	37.96	37.04
EBAY	41.67	38.43
FB	43.06	38.89
GOOG	41.20	37.96
INTC	47.22	41.20
MSFT	43.52	40.74
MU	46.30	38.89
PCAR	44.44	38.89
SMH	48.61	34.26
VOD	39.81	35.65

References

1. Almgren, Robert, and Neil Chriss. 2001. Optimal execution of portfolio transactions. *Journal of Risk* 3: 5–40.
2. Ashrafyan, Yuri, Tigran Bakaryan, Diogo Gomes, and Julian Gutierrez. A duality approach to a price formation MFG model. arXiv:2109.01791.
3. Bayraktar, Erhan, Ulrich Horst, and Ronnie Sircar. 2007. Queuing theoretic approaches to financial price fluctuations. *Handbooks in Operations Research and Management Science* 15: 637–677.
4. Bertsimas, Dimitris, and Andrew W. Lo. 1998. Optimal control of execution costs. *Journal of Financial Markets* 1 (1): 1–50.
5. Booth, G. Geoffrey, Raymond W. So, and Yiuman Tse. 1999. Price discovery in the german equity index derivatives markets. *Journal of Futures Markets: Futures, Options, and Other Derivative Products* 19 (6): 619–643.
6. Bouchaud, Jean-Philippe., Marc Mézard, Marc Potters, et al. 2002. Statistical properties of stock order books: empirical results and models. *Quantitative finance* 2 (4): 251–256.

7. Bouchaud, Jean-Philippe., Yuval Gefen, Marc Potters, and Matthieu Wyart. 2003. Fluctuations and response in financial markets: thesubtle nature ofrandom'price changes. *Quantitative finance* 4 (2): 176.
8. Brennan, Michael J., and Avanidhar Subrahmanyam. 1995. Investment analysis and price formation in securities markets. *Journal of Financial Economics* 38 (3): 361–381.
9. Brennan, Michael J., and Avanidhar Subrahmanyam. 1996. Market microstructure and asset pricing: On the compensation for illiquidity in stock returns. *Journal of Financial Economics*, 41 (3): 441–464.
10. Cardaliaguet, Pierre. 2010. *Notes on mean field games*. Technical report: Technical report.
11. Carmona, René, Daniel Lacker, et al. 2015. A probabilistic weak formulation of mean field games and applications. *The Annals of Applied Probability* 25 (3): 1189–1231.
12. Cartea, Álvaro, Sebastian Jaimungal, and José Penalva. 2015. *Algorithmic and high-frequency trading*. Cambridge University Press.
13. Chakraborti, Anirban, Ioane Muni Toke, Marco Patriarca, and Frédéric Abergel. 2011a. Econophysics review: II. Agent-based models. *Quantitative Finance* 11 (7): 1013–1041.
14. Chakraborti, Anirban, Ioane Muni Toke, Marco Patriarca, and Frédéric Abergel. 2011b. Econophysics review: I. Empirical facts. *Quantitative Finance* 11 (7): 991–1012.
15. Chu, Quentin C., Wen-liang Gideon Hsieh, and Yiuman Tse. 1999. Price discovery on the S&P 500 index markets: An analysis of spot index, index futures, and SPDRs. *International Review of Financial Analysis* 8 (1): 21–34.
16. Cohen,Kalman J., Steven F. Maier, Robert A. Schwartz, and David K. Whitcomb. 1981. Transaction costs, order placement strategy, and existence of the bid-ask spread. *Journal of Political Economy* 89 (2): 287–305.
17. Cont, Rama, Arseniy Kukanov, and Sasha Stoikov. 2014. The price impact of order book events. *Journal of Financial Econometrics* 12 (1): 47–88.
18. De Jong, Frank, and Peter C. Schotman. 2010. Price discovery in fragmented markets. *Journal of Financial Econometrics* 8 (1): 1–28.
19. Dias, Gustavo F., Marcelo Fernandes, and Cristina M. Scherrer. 2016. Price discovery in a continuous-time setting. *Journal of Financial Econometrics*.
20. Fruet Dias, Gustavo, Marcelo Fernandes, and Cristina Scherrer. 2018. Price discovery and market microstructure noise. Technical report, Working paper, Sao Paulo School of Economics.
21. Duffie, Darrell, and Haoxiang Zhu. 2017. Size discovery. *The Review of Financial Studies* 30 (4): 1095–1150.
22. Easley, David, and Maureen O'Hara. 1995. Market microstructure. *Handbooks in Operations Research and Management Science* 9: 357–383.
23. Evangelista, David, Yuri Saporito, and Yuri Thamsten. 2024. Price formation in financialmarkets: a finite player perspective. *Preprint*.
24. Fernandes, Marcelo, and Cristina M Scherrer. 2018. Price discovery in dual-class shares across multiple markets. *Journal of Futures Markets* 38 (1): 129–155.
25. Garbade, Kenneth D., and William L. Silber. 1979. Structural organization of secondary markets: Clearing frequency, dealer activity and liquidity risk. *The Journal of Finance* 34 (3): 577–593.
26. Gatheral, Jim. 2010. No-dynamic-arbitrage and market impact. *Quantitative finance* 10 (7): 749–759.
27. Gatheral, Jim, Alexander Schied, and Alla Slynko. 2012. Transient linear price impact and Fredholm integral equations. *Mathematical Finance: An International Journal of Mathematics, Statistics and Financial Economics* 22 (3): 445–474.
28. Gomes, Diogo, Julian Gutierrez, and Ricardo Ribeiro. 2023. A random-supply mean field game price model. *SIAM Journal on Financial Mathematics* 14 (1): 188–222.
29. Gomes, Diogo A., et al. 2020. A mean-field game approach to price formation. *Dynamic Games and Applications* 1–25.
30. Gonzalo, Jesus, and Clive Granger. 1995. Estimation of common long-memory components in cointegrated systems. *Journal of Business & Economic Statistics* 13 (1): 27–35.

31. Gonzalo, Jesus, and Serena Ng. 2001. A systematic framework for analyzing the dynamic effects of permanent and transitory shocks. *Journal of Economic Dynamics and Control* 25 (10): 1527–1546.
32. Grammig, Joachim, Michael Melvin, and Christian Schlag. 2005. Internationally cross-listed stock prices during overlapping trading hours: price discovery and exchange rate effects. *Journal of Empirical Finance* 12 (1): 139–164.
33. Guéant, Olivier. 2016. *The financial mathematics of market liquidity: From optimal execution to market making*, vol. 33. CRC Press.
34. Hasbrouck, Joel. 1995. One security, many markets: Determining the contributions to price discovery. *The Journal Of Finance* 50 (4): 1175–1199.
35. Ho, Thomas, and Hans R. Stoll. 1980. On dealer markets under competition. *The Journal of Finance* 35 (2): 259–267.
36. Ho, Thomas, and Hans R. Stoll. 1981. Optimal dealer pricing under transactions and return uncertainty. *Journal of Financial Economics* 9 (1): 47–73.
37. Ho, Thomas SY., and Hans R. Stoll. 1983. The dynamics of dealer markets under competition. *The Journal of Finance* 38 (4): 1053–1074.
38. Huang, M., R. P. Malhamé, and P. E. Caines. 2006. Large population stochastic dynamic games: closed-loop McKean-Vlasov systems and the Nash certainty equivalence principle. *Communications in Information and Systems* 6 (3): 221–251. ISSN 1526-7555. http://projecteuclid.org/getRecord?id=euclid.cis/1183728987.
39. Huang, M., P. E. Caines, and R. P. Malhamé. 2007. Large-population cost-coupled LQG problems with nonuniform agents: Individual-mass behavior and decentralized ϵ-Nash equilibria. *IEEE Transactions on Automatic Control* 52 (9): 1560–1571. ISSN 0018-9286. https://doi.org/10.1109/TAC.2007.904450
40. Kyle, Albert S. 1985. Continuous auctions and insider trading. *Econometrica: Journal of the Econometric Society* 53 (6): 1315–1335. https://doi.org/10.2307/1913210.
41. Boon Leong Lan and Ying Oon Tan. 2007. Statistical properties of stock market indices of different economies. *Physica A: Statistical Mechanics and its Applications* 375 (2): 605–611.
42. Lasry, J.-M., and P.-L. Lions. 2006a. Jeux à champ moyen. I. Le cas stationnaire. *Comptes Rendus Mathématique Académie des Sciences Paris* 343 (9): 619–625. ISSN 1631-073X.
43. Lasry, J.-M., and P.-L. Lions. 2006b. Jeux à champ moyen. II. Horizon fini et contrôle optimal. *Comptes Rendus Mathématique Académie des Sciences Paris* 343 (10): 679–684. ISSN 1631-073X.
44. Lasry, J.-M., and P.-L. Lions. 2007. Mean field games. *Japanese Journal of Mathematics* 2 (1): 229–260. ISSN 0289-2316.
45. Lehalle, Charles-Albert. 2013. Market microstructure knowledge needed for controlling an intra-day trading process. arXiv:1302.4592.
46. Lien, Donald, and Keshab Shrestha. 2009. A new information share measure. *Journal of Futures Markets: Futures, Options, and Other Derivative Products* 29 (4): 377–395.
47. Locke, Peter, and Zhan Onayev. 2007. Order flow, dealer profitability, and price formation. *Journal of Financial Economics* 85 (3): 857–887.
48. Obizhaeva, Anna A., and Jiang Wang. 2013. Optimal trading strategy and supply/demand dynamics. *Journal of Financial Markets* 16 (1): 1–32.
49. Pham, Huyên. 2009. *Continuous-time stochastic control and optimization with financial applications*, vol. 61. Springer Science & Business Media.
50. Potters, Marc, and Jean-Philippe. Bouchaud. 2003. More statistical properties of order books and price impact. *Physica A: Statistical Mechanics and its Applications* 324 (1–2): 133–140.
51. Shrivats, Arvind V., Dena Firoozi, and Sebastian Jaimungal. 2022. A mean-field game approach to equilibrium pricing in solar renewable energy certificate markets. *Mathematical Finance*. https://doi.org/10.1111/mafi.12345.

An Introduction to Monotonicity Methods in Mean-Field Games

Rita Ferreira, Diogo Gomes, and Teruo Tada

Abstract This chapter examines monotonicity techniques in the theory of mean-field games (MFGs). Originally, monotonicity ideas were used to establish the uniqueness of solutions for MFGs. Later, monotonicity methods and monotone operators were further exploited to build numerical methods and to construct weak solutions under mild assumptions. Here, after a brief discussion on the mean-field game formulation, we introduce the Minty method and regularization strategies for PDEs. These are then used to address typical stationary and time-dependent monotone MFGs and to establish the existence of weak solutions for such MFGs.

Keywords Mean-field games · Weak solutions · Stationary problems · Time-dependent problems · Monotone operators

MSC (2010) 35J56 · 35A01

1 Introduction

MFGs model the limit of differential games with a large population as the number of agents tends to infinity. In these models, each agent is rational and optimizes a cost functional. MFGs were motivated by questions in economics and engineering, and

R. Ferreira, D. Gomes, and T. Tada were partially supported by baseline and start-up funds from King Abdullah University of Science and Technology (KAUST) OSR-CRG2017-3452 and OSR-CRG2021-4674.

R. Ferreira · D. Gomes (✉) · T. Tada
CEMSE Division, King Abdullah University of Science and Technology (KAUST), Thuwal, Saudi Arabia
e-mail: diogo.gomes@kaust.edu.sa

R. Ferreira
e-mail: rita.ferreira@kaust.edu.sa

T. Tada
e-mail: teruo.tada@kaust.edu.sa

E. A. Pimentel and B. Toni (eds.), *Differential and Algorithmic Intelligent Game Theory*, STEAM-H: Science, Technology, Engineering, Agriculture, Mathematics & Health,
https://doi.org/10.1007/978-3-031-97733-6_4

were introduced by Lasry and Lions in [42–44] and Caines, Huang, and Malhamé in [40, 41]. Often, these games comprise a system of two equations, a Hamilton–Jacobi equation and a Fokker–Planck equation, which can be associated with a monotone operator. This structure is key to establish the uniqueness of solutions, as used in [42]. Here, we illustrate further applications and techniques. In particular, we address how to build weak solutions for stationary and time-dependent MFGs.

A main difficulty in the MFG theory is to establish the existence of solutions. While Hamilton–Jacobi equations and Fokker–Planck equations are extremely well understood, the coupling between these two equations presents substantial difficulties. Monotone operators ideas give a unified approach to address a wide class of MFGs. This approach, which we illustrate here, was first introduced in [26] and was motivated by prior works in the numerical approximation of MFGs in [2]. Subsequently, it was further developed in [27] for stationary problems with boundary conditions, in [28] for time-dependent problems, and in [4] for the planning problem with congestion. The purpose of the present chapter is to describe in a simplified setting the main ideas on those references.

In recent years, there has been much progress on the existence and regularity of solutions for MFGs. Most of the available literature relies heavily on the structure of the specific problem under consideration and often do not generalize in a straightforward way to other problems. A historical account of part of the theory can be found in the lectures [45] and is available in the books and surveys [1, 6, 12, 16, 30, 37, 39]. Partial differential equation methods have played an essential role in the development of MFGs from the beginning, see [42, 43]. Subsequently, several authors have looked into core matters such as the existence and regularity of solutions. Smooth solutions were obtained in [32, 38, 48, 53] for the stationary case and in [19, 33–36] for the time-dependent case. These results were obtained by combining the non-linear adjoint method to obtain a priori estimates with the continuation method. Around the same time, the theory of weak solutions saw substantial progress in the elliptic case in [7, 47] and in the parabolic case in [14, 49–51]. Degenerate elliptic or parabolic problems were examined in [14] and, in the particular hypoelliptic case, were studied in [23, 25, 46]. First-order MFGs present substantial challenges and were examined in [13, 15, 29]. The theory for MFGs for problems with boundary conditions and state constraints, especially for first-order MFGs, is not yet complete. The case where there is an invariance condition for the state space was investigated in [52]. The authors in [8–11] use Lagrangian methods to investigate MFGs with state constraints. This approach relies on a weak notion of MFG equilibrium defined in a Lagrangian setting rather than with a PDE system. In [5, 8, 46], the equilibria are probability measures defined on a set of admissible trajectories. Once the existence of a relaxed MFG equilibrium is ensured, we can investigate the regularity of solutions and give meaning to the PDE system and the related boundary conditions [9, 10].

While our focus here are monotone MFGs, this introduction would not be complete without mentioning some of the research on non-monotone MFGs. These were studied in particular one-dimensional cases in [31]. Then, using PDE methods, [17, 20, 21] prove the existence under suitable growth assumptions for the problem and in [18] for the short-time problem. Finally, the short-time case or small data was examined in [3] based on the analysis of the Fourier coefficients of the solutions, a technique derived from prior works in fluid mechanics.

A model stationary MFG is the following. Agents' state is the d-dimensional torus, $\mathbb{T}^d$. The distribution of agents is given by a density $m : \mathbb{T}^d \to \mathbb{R}_0^+$, not necessarily normalized, where $\mathbb{R}_0^+ = [0, +\infty)$. Each agent selects a trajectory $\mathbf{x} : \mathbb{R}_0^+ \to \mathbb{T}^d$ and seeks to minimize the discounted cost

$$\int_0^{+\infty} e^{-t}(L(\dot{\mathbf{x}}(t)) - V(\mathbf{x}(t)) + g(m(\mathbf{x}(t))))\,dt.$$

The discount term e^{-t} accounts for agents exiting the game at a random time with an exponential constant rate. These exiting agents are replaced by agents that begin their trajectories at a random point in $\mathbb{T}^d$ chosen according to a rate function $\phi \geqslant 0$. The Lagrangian $L : \mathbb{R}^d \to \mathbb{R}$ is a uniformly convex function (or, alternatively, strictly convex and coercive) that encodes the cost of moving at the speed $\dot{\mathbf{x}}(t)$. The potential $V : \mathbb{T}^d \to \mathbb{R}$ represents the spatial preferences of the agents. Finally, $g : \mathbb{R}_0^+ \to \mathbb{R}$ determines congestion effects and is where the interaction between each agent and the population is encoded. Let

$$H(p) = \sup_{v \in \mathbb{R}^d}(-v \cdot p - L(v)), \quad p \in \mathbb{R}^d, \tag{1}$$

be the Legendre transform of L. The value function,

$$u(x) = \inf_{\substack{\mathbf{x} \in W^{1,\infty}(0,+\infty) \\ \mathbf{x}(0)=x}} \int_0^{+\infty} e^{-t}(L(\dot{\mathbf{x}}(t)) - V(\mathbf{x}(t)) + g(m(\mathbf{x}(t))))\,dt,$$

if regular enough, it solves the Hamilton–Jacobi equation

$$u + H(Du) - g(m) + V = 0.$$

Further, the optimal trajectory is given in feedback form by

$$\dot{\mathbf{x}} = -D_p H(Du(\mathbf{x})).$$

In this model, agents are rational and therefore the whole population follows the preceding ODE. This gives rise to the transport equation

$$m - \operatorname{div}\left(m D_p H(Du)\right) - \phi = 0$$

that determines m. Gathering the Hamilton–Jacobi and the transport equation, we obtain the MFG system

$$\begin{cases} -u - H(Du) + g(m) - V = 0 \text{ in } \mathbb{T}^d, \\ m - \operatorname{div}\left(m D_p H(Du)\right) - \phi = 0 \text{ in } \mathbb{T}^d. \end{cases} \tag{2}$$

For more details on the derivation of MFG models, including the time-dependent case, we refer the reader to [37]. The preceding system can be associated with the operator

$$F\begin{bmatrix} w \\ v \end{bmatrix} = \begin{bmatrix} -v - H(Dv) + g(w) - V \\ w - \operatorname{div}(wD_pH(Dv)) - \phi \end{bmatrix}. \tag{3}$$

If g is increasing and H is convex, F is monotone on its domain $D \subset L^2(\mathbb{T}^d; \mathbb{R}_0^+) \times L^2(\mathbb{T}^d)$. As we mentioned before, the focus of this chapter is to explore this monotonicity.

A first application of monotonicity is the uniqueness of solutions to MFGs, which is discussed in Sect. 2. Concerning existence of solutions, we work with weak solutions. Before addressing MFGs, for pedgagogical reasons, we illustrate the main techniques on simpler PDEs in Sect. 3. There, we define weak solutions for monotone operators, discuss the Minty's method to obtain weak solutions, and illustrate two main technical points, regularization strategies and corresponding existence of solutions for the regularized problems. We then extend these ideas to address MFGs in Sects. 4 and 5.

2 Monotonicity and Uniqueness

Often, the equations comprising MFGs can be interpreted as a monotone operator, whose notion we recall next. Let H be a Hilbert space endowed with an inner product, $(\cdot, \cdot)$, and let $D(\cdot)$ be the domain of an operator on H. We say that an operator, $\mathfrak{F} : D(F) \subset H \to H$, is monotone if for every $u, v \in D(\mathfrak{F})$, $\mathfrak{F}$ satisfies

$$\big(\mathfrak{F}(u) - \mathfrak{F}(v), u - v\big) \geqslant 0. \tag{4}$$

We say that $\mathfrak{F}$ is strictly monotone if for all $u, v \in D(\mathfrak{F}) \subset H$ such that $u \not\equiv v$, we have

$$(\mathfrak{F}(u) - \mathfrak{F}(v), u - v) > 0. \tag{5}$$

If $\mathfrak{F}$ is strictly monotone, then there exists at most one solution to the problem of finding $u \in D(\mathfrak{F})$ for which $\mathfrak{F}(u) = 0$. In fact, if $\mathfrak{F}$ is strictly monotone and u, $v \in D(\mathfrak{F})$ are such that $\mathfrak{F}(u) = \mathfrak{F}(v) = 0$, then the left-hand side of (5) vanishes, which can only be if $u \equiv v$.

Proposition 2.1 *Consider the operator F given by (3) associated with the MFG in (2). Assume that H is convex, g is increasing, and let $D(F) = C^2(\mathbb{T}^d; \mathbb{R}_0^+) \times C^1(\mathbb{T}^d)$. Then, F is monotone in $L^2(\mathbb{T}^d) \times L^2(\mathbb{T}^d)$.*

Proof Let $\langle \cdot, \cdot \rangle$ denote the inner product in $L^2(\mathbb{T}^d) \times L^2(\mathbb{T}^d)$. Given (m_1, u_1), $(m_2, u_2) \in D(F)$, integrating by parts and using the assumptions on H and g yields

$$\begin{aligned}
&\left\langle F\begin{bmatrix} m_1 \\ u_1 \end{bmatrix} - F\begin{bmatrix} m_2 \\ u_2 \end{bmatrix}, \begin{bmatrix} m_1 \\ u_1 \end{bmatrix} - \begin{bmatrix} m_2 \\ u_2 \end{bmatrix}\right\rangle \\
&\quad = \int_{\mathbb{T}^d} m_1\big(H(Du_2) - H(Du_1) - D_pH(Du_1)\cdot(Du_2 - Du_1)\big)\,\mathrm{d}x \\
&\quad + \int_{\mathbb{T}^d} m_2\big(H(Du_1) - H(Du_2) - D_pH(Du_2)\cdot(Du_1 - Du_2)\big)\,\mathrm{d}x \\
&\quad + \int_{\mathbb{T}^d} (m_1 - m_2)(g(m_1) - g(m_2))\,\mathrm{d}x \geqslant 0,
\end{aligned} \tag{6}$$

which establishes the monotonicity of F. □

Remark 2.2 Under the assumptions of the preceding proposition, the following second-order stationay MFG

$$\begin{cases} u - \Delta u + H(Du) - g(m) - V(x) = 0 & \text{in } \mathbb{T}^d, \\ m - \Delta m - \operatorname{div}(mD_pH(Du)) - \phi(x) = 0 & \text{in } \mathbb{T}^d, \end{cases} \tag{7}$$

and time-dependent MFG

$$\begin{cases} -u_t - \Delta u + H(Du) - g(m) - V = 0 & \text{in } (0,T)\times\mathbb{T}^d, \\ m_t - \Delta m - \operatorname{div}(mD_pH(Du)) = 0 & \text{in } (0,T)\times\mathbb{T}^d, \end{cases} \tag{8}$$

with initial-terminal boundary conditions

$$m(0,\cdot) = m_0,\; u(T,\cdot) = u_T \quad \text{in } \mathbb{T}^d, \tag{9}$$

with $m_0 : \mathbb{T}^d \to \mathbb{R}_0^+$, $\int_{\mathbb{T}^d} m_0 = 1$, and $u_T : \mathbb{T}^d \to \mathbb{R}$, share the same monotonone structure. More precisely, define $F_1\colon\ D(F_1) \subset L^2(\mathbb{T}^d) \times L^2(\mathbb{T}^d) \to L^2(\mathbb{T}^d) \times L^2(\mathbb{T}^d)$ by

$$F_1\begin{bmatrix} m \\ u \end{bmatrix} = \begin{bmatrix} -u + \Delta u - H(Du) + g(m) + V \\ m - \Delta m - \operatorname{div}(mD_pH(Du)) - \phi \end{bmatrix}, \quad (m,u) \in D(F_1),$$

and $F_2\colon\ D(F_2) \subset L^2([0,T]\times\mathbb{T}^d) \times L^2([0,T]\times\mathbb{T}^d) \to L^2([0,T]\times\mathbb{T}^d) \times L^2([0,T]\times\mathbb{T}^d)$ by

$$F_2\begin{bmatrix} m \\ u \end{bmatrix} = \begin{bmatrix} u_t + \Delta u - H(Du) + g(m) + V \\ m_t - \Delta m - \operatorname{div}(mD_pH(Du)) \end{bmatrix}, \quad (m,u) \in D(F_2).$$

By selecting $D(F_1) = C^2(\mathbb{T}^d; \mathbb{R}_0^+) \times C^2(\mathbb{T}^d)$ and

$$\begin{aligned} D(F_2) = \{&m \in C^2([0,T]\times\mathbb{T}^d; \mathbb{R}_0^+) : m(x,0) = m_0\} \\ &\times \{u \in C^2([0,T]\times\mathbb{T}^d) : u(x,T) = u_T\} \end{aligned}$$

then, following the steps in the proof of Proposition 2.1, we obtain the monotonicity of F_1 and F_2 in $L^2(\mathbb{T}^d) \times L^2(\mathbb{T}^d)$ and $L^2([0,T] \times \mathbb{T}^d) \times L^2([0,T] \times \mathbb{T}^d)$, respectively.

The previous operators are not strictly monotone, even if H is strictly convex and g is strictly increasing; the lack of strict monotonicity arises from the possibility of m vanishing, thus (6) could be identically 0 without $Du_1 = Du_2$. Thus, to prove uniqueness of solutions, we either need to add further conditions on those solutions or provide additional arguments that give the strict positivity of m.

Proposition 2.3 *Consider the operator F given by (3) associated with the MFG in (2). Assume that H is strictly convex, g is strictly increasing, and $\phi > 0$ in $\mathbb{T}^d$. Then, there is at most one solution $(m,u) \in C^2(\mathbb{T}^d;\mathbb{R}_0^+) \times C^1(\mathbb{T}^d)$ to (2).*

Proof Let $(m,u) \in C^2(\mathbb{T}^d;\mathbb{R}_0^+) \times C^1(\mathbb{T}^d)$ solve (2). We first prove that $m > 0$. Assume by contradiction that $\min_{x\in\mathbb{T}^d} m = m(x^*) = 0$ for some $x^* \in \mathbb{T}^d$. Then, $\operatorname{div}(m(x^*)D_pH(Du(x^*))) = 0$ since both m and Dm vanish at x^*. Thus

$$0 = m(x^*) - \operatorname{div}(m(x^*)D_pH(Du(x^*))) - \phi(x^*) = -\phi(x^*) < 0,$$

which is a contradiction. Hence, m is strictly positive in $\mathbb{T}^d$.

To prove uniqueness, let (m_1,u_2) and (m_2,u_2) be solutions to (2) in $C^2(\mathbb{T}^d;\mathbb{R}_0^+) \times C^1(\mathbb{T}^d)$. Then m_1 and m_2 are strictly positive. Consequently, (6), combined with the convexity of H and the monotonicity of g, gives

$$\begin{aligned}
&H(Du_2) - H(Du_1) - D_pH(Du_1)\cdot(Du_2 - Du_1) = 0,\\
&H(Du_1) - H(Du_2) - D_pH(Du_2)\cdot(Du_1 - Du_2) = 0,\\
&(m_1 - m_2)(g(m_1) - g(m_2)) = 0.
\end{aligned}$$

Therefore, the strict convexity of H gives $Du_1 = Du_2$ and the strict monotonicity of g gives $m_1 = m_2$. Finally, the Hamilton–Jacobi equation combined with these facts gives $u_1 = u_2$. □

Remark 2.4 The uniqueness for the second-order stationary MFG in (7) is handled similarly. For the second-order time-dependent case, the argument must be modified slightly: we require $m_0 > 0$ and use the parabolic maximum principle to get $m > 0$.

3 Minty's Method and Regularization of Monotone Operators

Before considering MFGs, we focus on monotone operators and explain Minty's method for studying the kernel of such operators. The main idea of this method is to construct a convenient monotone approximation for which the kernel is simpler to study. Then, using the monotonicity and asymptotic analysis, we build a weak solution for the original problem. We refer the reader to [24] for a detailed description of Minty's method, where this method is used for finding weak solutions for nonlinear PDEs.

Let H be a Hilbert space and assume that $F\colon D(F) \subset H \to H$ is a monotone operator (see (4)). The study of the kernel of F is often formulated as a (strong) variational inequality: find $u \in D(F)$ such that for all $v \in D(F)$, we have

$$(F(u), u - v) \leqslant 0. \tag{10}$$

Note that if $D(F) = H$, then taking $v = u \pm w$ for an arbitrary $w \in H$, we conclude that (10) is equivalent to finding $u \in H$ such that $F(u) = 0$.

In many instances, the (strong) variational inequality is either hard to address or does not have solutions, which motivates the so-called weak variational inequality associated with F. This weak variational inequality consists of finding $u \in H$ such that for all $v \in D(F)$, we have

$$(F(v), u - v) \leqslant 0. \tag{11}$$

A weak solution to (10) is an element $u \in H$ satisfying (11). Because of the monotonicity of F, any (strong) solution to (10) is also a weak solution to (10). Conversely, if F is continuous and $D(F)$ is convex, then weak solutions belonging to $D(F)$ are also strong solutions. In fact, let $u \in D(F)$ be a weak solution to (10), let $\lambda \in (0, 1)$, fix $\bar{v} \in D(F)$, and set $v = \lambda u + (1 - \lambda)\bar{v} \in D(F)$. Then, by (11), we have

$$0 \geqslant \big(F(\lambda u + (1 - \lambda)\bar{v}), (1 - \lambda)(u - \bar{v})\big).$$

Dividing the preceding inequality by $1 - \lambda$, and then letting $\lambda \to 1^-$, we obtain

$$0 \geqslant (F(u), u - \bar{v})$$

by the continuity of F. Because $\bar{v} \in D(F)$ is arbitrary, we conclude that (10) holds; that is, u is also a strong solution to (10).

A common method to finding weak solutions to (10) is Minty's method, which is hinged on finding a convenient monotone approximation of F. More precisely, assume that there exists a dense subset, $D \subset D(F)$, and, for each $k \in \mathbb{N}$, there exists an operator, $F_k\colon D \subset H \to H$, satisfying the following conditions:

1. *(monotonicity)* for all $v, u \in D$, we have

$$(F_k(u) - F_k(v), u - v) \geqslant 0; \tag{12}$$

2. *(convergence to F)* for all $v \in D$, we have

$$F_k(v) \to F(v) \ \text{ strongly in } H; \tag{13}$$

3. *(lower semicontinuity-type condition)* for all $v \in D(F)$, there exists $\{v_k\}_{k=1}^{\infty} \subset D$ such that

$$v_k \to v \text{ in } H \text{ and } F(v_k) \rightharpoonup F(v) \text{ weakly in } H; \tag{14}$$

4. *(strong solutions)* there exists $u_k \in D$ such that for all $v \in D$, we have

$$(F_k(u_k), u_k - v) \leqslant 0; \tag{15}$$

5. *(compactness)* given $\{u_k\}_{k=1}^{\infty}$ as in (15), there exists $\bar{u} \in H$ such that, extracting a subsequence if necessary, we have

$$u_k \rightharpoonup \bar{u} \text{ weakly in } H. \tag{16}$$

Proposition 3.1 (Minty's method) *Let F be a monotone operator in a Hilbert space H and F_k be as above satisfying conditions 1–5. Then, any $\bar{u} \in H$ given by (16) is a weak solution to (10).*

Proof By combining (12) with u replaced by u_k and (15), we obtain that $(F_k(v), u_k - v) \leqslant 0$ for all $v \in D$. Letting $k \to \infty$ in this estimate, (13) and (16) yield

$$(F(v), \bar{u} - v) \leqslant 0 \tag{17}$$

for all $v \in D$. Finally, using (14), we conclude that (17) holds for all $v \in D(F)$, which shows that $\bar{u}$ is a weak solution to (10). □

In the following subsections, we illustrate how to construct such monotone approximations, F_k.

3.1 Regularization Strategies

We now show how to construct monotone approximations as mentioned above. We will use a regularization method based on higher-order perturbations. Rather than addressing MFG models, we consider a simpler linear operator to illustrate the main ideas. In the next section, we examine operators that arise in MFGs.

To simplify, assume that the domain is the d-dimensional torus, $\mathbb{T}^d$, and fix $\mathbf{b} \in C^1(\mathbb{T}^d; \mathbb{R}^d)$ with $\mathrm{div}(\mathbf{b}) = 0$ and $f \in C^1(\mathbb{T}^d)$ with $f \not\equiv 0$. Let $F : H^1(\mathbb{T}^d) \subset L^2(\mathbb{T}^d) \to L^2(\mathbb{T}^d)$ be defined by

$$F(u) = u - \mathbf{b} \cdot Du - f \tag{18}$$

for $u \in H^1(\mathbb{T}^d)$. Consider the problem of finding $u \in H^1(\mathbb{T}^d)$ such that $F(u) = 0$. Observe that $u \equiv 0$ is not a solution to this problem because $f \not\equiv 0$. Moreover, we cannot apply the Lax–Milgram theorem directly to solve (18) because the bilinear form associated with (18) is neither continuous in $L^2(\mathbb{T}^d)$ nor coercive in $H^1(\mathbb{T}^d)$. Instead, we use Minty's method as described above, for which we start by establishing the monotonicity of F.

Proposition 3.2 *The operator F given by (18) is monotone in $L^2(\mathbb{T}^d)$.*

Proof Given $u, v \in H^1(\mathbb{T}^d)$, an integration by parts and Proposition 3.2 yield

$$
\begin{aligned}
&(F(u) - F(v), u - v)\\
&= \int_{\mathbb{T}^d} \left((u-v)^2 - \mathbf{b}\cdot D(u-v)(u-v)\right) \mathrm{d}x = \int_{\mathbb{T}^d} \left((u-v)^2 - \frac{1}{2}\mathbf{b}\cdot D(u-v)^2\right) \mathrm{d}x\\
&= \int_{\mathbb{T}^d} \left((u-v)^2 + \frac{1}{2}\operatorname{div}(\mathbf{b})(u-v)^2\right) \mathrm{d}x = \int_{\mathbb{T}^d} (u-v)^2 \,\mathrm{d}x \geqslant 0.
\end{aligned}
$$

□

To use Minty's method, we must construct an approximation of F that preserves monotonicity. One possibility is to choose, for $\epsilon > 0$, the regularized operator

$$
F_\epsilon(u) = F(u) + \epsilon(u + \Delta^2 u) \tag{19}
$$

with domain $D = H^4(\mathbb{T}^d)$.

Proposition 3.3 *The operator F_ϵ given by (19) satisfies conditions (12), (13), (14), and (16).*

Proof For all $u, v \in H^4(\mathbb{T}^d)$, an integration by parts yields

$$
\begin{aligned}
&(F_\epsilon(u) - F_\epsilon(v), u - v)\\
&\quad = (F(u) - F(v), u - v) + \epsilon\left(\|u - v\|^2_{L^2(\mathbb{T}^d)} + \|\Delta u - \Delta v\|^2_{L^2(\mathbb{T}^d)}\right) \geqslant 0.
\end{aligned}
$$

Thus, F_ϵ is (strictly) monotone over $L^2(\mathbb{T}^d)$. Hence, (12) holds.

Because $\epsilon(v + \Delta^2 v) \to 0$ strongly in L^2 for $v \in H^4(\mathbb{T}^d)$, (13) also holds.

Next, by density of $H^4(\mathbb{T}^d)$ in $H^1(\mathbb{T}^d)$, for any $v \in H^1(\mathbb{T}^d)$, there exists a sequence $\{v_k\}_k$ in $H^4(\mathbb{T}^d)$ that converges strongly in $H^1(\mathbb{T}^d)$ to v. Hence, $v_k \to v$ in $L^2(\mathbb{T}^d)$ and $F(v_k) \rightharpoonup F(v)$ weakly in $L^2(\mathbb{T}^d)$. Thus, (14) holds.

Assume that there exists $u_\epsilon \in H^4(\mathbb{T}^d)$ such that

$$
F_\epsilon(u_\epsilon) = 0 \quad \text{in } \mathbb{T}^d. \tag{20}
$$

Then, multiplying both sides of (20) by u_ϵ, integrating by parts, and using the condition $\operatorname{div}(\mathbf{b}) = 0$, we get

$$
0 = \int_{\mathbb{T}^d} \left(F(u_\epsilon)u_\epsilon + \epsilon\left(u_\epsilon^2 + (\Delta u_\epsilon)^2\right)\right) \mathrm{d}x = \int_{\mathbb{T}^d} \left(u_\epsilon^2 + f(x)u_\epsilon + \epsilon\left(u_\epsilon^2 + (\Delta u_\epsilon)^2\right)\right) \mathrm{d}x.
$$

Applying Young's inequality, we conclude that there exists a positive constant, C, such that

$$
\int_{\mathbb{T}^d} \left(u_\epsilon^2 + \epsilon\left(u_\epsilon^2 + (\Delta u_\epsilon)^2\right)\right) \mathrm{d}x \leqslant C.
$$

Because C is independent of ϵ, we obtain uniform a priori estimates in $H^2(\mathbb{T}^d)$ for the solutions u_ϵ, from which (16) holds by the Rellich–Kondrachov theorem. □

Proposition 3.4 *The operator F_ϵ given by (19) satisfies condition (15).*

In the subsequent sections, we describe three approaches to prove the preceding propositon that establishes the existence of solutions to the regularized problem. These are a variational approach, a bilinear form approach, and a continuity method approach.

Variational Problem Approach

Given $u \in H^1(\mathbb{T}^d)$, we define a functional, $G_u\colon H^2(\mathbb{T}^d) \to \mathbb{R}$, by

$$G_u(v) = \int_{\mathbb{T}^d} \left(\frac{\epsilon}{2}(v^2 + (\Delta v)^2) + F(u)v\right) \mathrm{d}x, \quad v \in H^2(\mathbb{T}^d), \tag{21}$$

where F is given by (18).

Proposition 3.5 *Fix $u_0 \in H^1(\mathbb{T}^d)$ and set $G = G_{u_0}$. The variational problem of finding $\bar{v} \in H^2(\mathbb{T}^d)$ satisfying*

$$G(\bar{v}) = \inf_{v \in H^2(\mathbb{T}^d)} G(v) \tag{22}$$

has a unique minimizer.

Proof The functional G is weakly lower-semicontinuous in $H^2(\mathbb{T}^d)$. To apply the direct method in the calculus of variations, it suffices to show that it is coercive. This follows from the inequality

$$\begin{aligned} G(v) &= \int_{\mathbb{T}^d} \left(\frac{\epsilon}{2}(v^2 + (\Delta v)^2) + F(u_0)v\right) \mathrm{d}x \geqslant \int_{\mathbb{T}^d} \left(\frac{\epsilon}{2}(v^2 + (\Delta v)^2) - C|v|\right) \mathrm{d}x \\ &\geqslant \int_{\mathbb{T}^d} \left(\frac{\epsilon}{2}(v^2 + (\Delta v)^2) - \frac{\epsilon}{4}v^2 - \frac{4C}{\epsilon}\right) \mathrm{d}x \geqslant C_\epsilon \|v\|^2_{H^2(\mathbb{T}^d)} - \frac{C}{\epsilon}, \end{aligned}$$

which is a consequence of Young's and the Gagliardo–Nirenberg interpolation inequalities. Because G is strictly convex, the minimizer is unique. □

The preceding proposition enable us to define an operator $\mathcal{U}\colon H^1(\mathbb{T}^d) \to H^1(\mathbb{T}^d)$ by

$$\mathcal{U}[u_0] = \bar{v}, \tag{23}$$

where $u_0 \in H^1(\mathbb{T}^d)$ and $\bar{v}$ is the solution of (22) given by Proposition 3.5. To prove Proposition 3.4, we aim at showing that $\mathcal{U}$ has a fixed point. Note that the range of $\mathcal{U}$ is $H^2(\mathbb{T}^d)$, hence any fixed point of $\mathcal{U}$ is automatically in $H^2(\mathbb{T}^d)$. Once the existence of a fixed point is established, it satisfies (15) by the following proposition.

Proposition 3.6 *Let $u \in H^2(\mathbb{T}^d)$ be a fixed point of $\mathcal{U}$. Then,*

$$\int_{\mathbb{T}^d} \Big(\epsilon\big(u(\omega - u) + \Delta u \Delta(\omega - u)\big) + F(u)(\omega - u)\Big)\, \mathrm{d}x \geqslant 0 \tag{24}$$

for all $\omega \in H^2(\mathbb{T}^d)$.

Proof Fix $\omega \in H^2(\mathbb{T}^d)$ and let $I\colon [0,1] \to \mathbb{R}$ be given by

$$I(t) = G(u + t(\omega - u)), \quad t \in [0,1].$$

Because u is a minimizer, we have $I(0) \leqslant I(t)$. Consequently,

$$0 \leqslant \frac{1}{t}(I(t) - I(0)) = \int_{\mathbb{T}^d} \Big(\epsilon\big(u(\omega - u) + \Delta u \Delta(\omega - u)\big) + F(u)(\omega - u)\Big)\, \mathrm{d}x + \mathcal{O}(t)$$

for all $t \in (0,1]$. Letting $t \to 0$ in the preceding inequality, we conclude that u satisfies (24). □

Remark 3.7 Arguing as in the previous proposition for $\bar{v} = \mathcal{U}(u_0)$, we obtain that

$$\int_{\mathbb{T}^d} \Big(\epsilon\big(\bar{v}(\omega - \bar{v}) + \Delta \bar{v} \Delta(\omega - \bar{v})\big) + F(u_0)(\omega - \bar{v})\Big)\, \mathrm{d}x \geqslant 0.$$

Further, by taking $t \in [-1, 0]$ in the preceding proof gives

$$\int_{\mathbb{T}^d} \Big(\epsilon\big(\bar{v}(\omega - \bar{v}) + \Delta \bar{v} \Delta(\omega - \bar{v})\big) + F(u_0)(\omega - \bar{v})\Big)\, \mathrm{d}x = 0,$$

which is the weak form of the Euler–Lagrange equation for the functional G. In particular, any fixed point of $\mathcal{U}$ in $H^2(\mathbb{T}^d)$ is a weak solution of

$$\epsilon(u + \Delta^2 u) + F(u) = 0.$$

Accordingly, $\Delta^2 u \in L^2(\mathbb{T}^d)$, which gives $u \in H^4(\mathbb{T}^d)$.

To prove that $\mathcal{U}$ has a fixed point, we use the following version of Schaefer's fixed-point theorem (see, for instance, [27, Theorem 6.2] for its proof).

Theorem 3.8 *Let $\mathcal{X}$ be a closed convex subset of a Banach space such that $0 \in \mathcal{X}$. Assume that $\mathcal{U}\colon \mathcal{X} \to \mathcal{X}$ is continuous and compact. Also, assume that the set*

$$\big\{w \in \mathcal{X} \mid w = \lambda \mathcal{U}[w] \text{ for some } \lambda \in [0,1]\big\}$$

is bounded. Then, there exists a fixed point, $w \in \mathcal{X}$, such that $w = \mathcal{U}[w]$.

To apply the preceding theorem, we first prove the continuity and compactness of $\mathcal{U}$.

Proposition 3.9 *The operator $\mathcal{U}$ in (23) is continuous and compact in $H^1(\mathbb{T}^d)$.*

Proof First, we establish continuity. Let $\{u_n\}_{n=1}^{\infty} \subset H^1(\mathbb{T}^d)$ and $u \in H^1(\mathbb{T}^d)$ be such that $u_n \to u$ in $H^1(\mathbb{T}^d)$. Recalling (21), set $G^* = G_u$ and $G_n = G_{u_n}$. Let

$$\bar{v} = \mathcal{U}[u], \quad \bar{v}_n = \mathcal{U}[u_n];$$

that is, $\bar{v}$ and $\bar{v}_n$ solve (22) with G replaced by G^* and G_n, respectively. Because $\bar{v}$ and $\bar{v}_n$ are minimizers over $H^2(\mathbb{T}^d)$ of G^* and G_n, respectively, we have

$$\begin{aligned} G^*(\bar{v}) + G_n(\bar{v}_n) &\leqslant G^*\Big(\frac{\bar{v}+\bar{v}_n}{2}\Big) + G_n\Big(\frac{\bar{v}+\bar{v}_n}{2}\Big) \\ &= \frac{1}{2}G^*(\bar{v}) + \frac{1}{2}G^*(\bar{v}_n) + \frac{1}{2}G_n(\bar{v}) + \frac{1}{2}G_n(\bar{v}_n) \\ &\quad - \int_{\mathbb{T}^d} \frac{\epsilon}{4}\Big[(\bar{v}-\bar{v}_n)^2 + (\Delta\bar{v} - \Delta\bar{v}_n)^2\Big]\,\mathrm{d}x, \end{aligned}$$

which can be written as

$$\int_{\mathbb{T}^d} \frac{\epsilon}{4}\Big[(\bar{v}-\bar{v}_n)^2 + (\Delta\bar{v} - \Delta\bar{v}_n)^2\Big]\,\mathrm{d}x \leqslant \frac{1}{2}\big(- G^*(\bar{v}) + G^*(\bar{v}_n) + G_n(\bar{v}) - G_n(\bar{v}_n)\big).$$

Using Young's inequality, we obtain that

$$\begin{aligned} \int_{\mathbb{T}^d} \frac{\epsilon}{4}\Big[(\bar{v}-\bar{v}_n)^2 + (\Delta\bar{v} - \Delta\bar{v}_n)^2\Big]\,\mathrm{d}x &\leqslant \int_{\mathbb{T}^d} |\bar{v}-\bar{v}_n||F(u) - F(u_n)|\,\mathrm{d}x \\ &\leqslant \int_{\mathbb{T}^d} \frac{\epsilon}{8}(\bar{v}-\bar{v}_n)^2\,\mathrm{d}x + \frac{C}{\epsilon}\|u-u_n\|^2_{H^1(\mathbb{T}^d)}. \end{aligned} \tag{25}$$

Hence, using the Gagliardo–Nirenberg interpolation inequality and the convergence $u_n \to u$ in $H^1(\mathbb{T}^d)$, we conclude from the previous estimate that $\bar{v}_n$ converges to $\bar{v}$ in $H^2(\mathbb{T}^d)$. This proves that $\mathcal{U}$ is continuous over $H^1(\mathbb{T}^d)$.

Next, we establish compactness. For that, let u_n be a bounded sequence in $H^1(\mathbb{T}^d)$. We claim that the corresponding sequence $\bar{v}_n = \mathcal{U}(u_n)$ is pre-compact. By choosing $u = 0$ and $\bar{v} = \mathcal{U}(0)$, inequality (25) combined with Gagliardo–Nirenberg estimate gives that $\bar{v}_n$ is bounded in $H^2(\mathbb{T}^d)$, hence pre-compact in $H^1(\mathbb{T}^d)$ by the Rellich–Kondrachov theorem. □

Proposition 3.10 *The operator $\mathcal{U}$ in (23) has a fixed point $\bar{u} \in H^1(\mathbb{T}^d) \cap H^2(\mathbb{T}^d)$.*

Proof To establish the existence of a fixed point, we apply Theorem 3.8. For that, fix $\lambda \in [0, 1]$ and assume that u_λ satisfies

$$u_\lambda = \lambda\mathcal{U}[u_\lambda].$$

If $\lambda = 0$, then $u_\lambda = 0$. If $\lambda \in (0, 1]$, then arguing as in Remark 3.7, we have

$$\int_{\mathbb{T}^d} \left(\epsilon\big(u_\lambda(\lambda\omega - u_\lambda) + \Delta^k u_\lambda \Delta^k(\lambda\omega - u_\lambda)\big) + \lambda F(u_\lambda)(\lambda\omega - u_\lambda)\right) \mathrm{d}x = 0$$

for all $\omega \in H^2(\mathbb{T}^d)$.

Setting $\omega = 0$, integrating by parts with $\operatorname{div}(\mathbf{b}) = 0$, and using Young's inequality, we get

$$\int_{\mathbb{T}^d} \left(\lambda u_\lambda^2 + \epsilon\big(u_\lambda^2 + (\Delta u_\lambda)^2\big)\right) \mathrm{d}x \leqslant C,$$

where C is independent of λ and ϵ. Thus, using the Gagliardo–Nirenberg interpolation inequality, u_λ is bounded uniformly for $\lambda \in [0, 1]$ in $H^2(\mathbb{T}^d)$ and hence in $H^1(\mathbb{T}^d)$. Therefore, by the compactness and continuity proven in Proposition 3.9 combined with Theorem 3.8, we conclude that $\mathcal{U}$ has a fixed point $\bar{u} \in H^1(\mathbb{T}^d) \cap H^2(\mathbb{T}^d)$. □

Proof of Proposition 3.4 Proposition 3.4 follows by combining the existence of a fixed point established in Proposition 3.10, its characterization in Proposition 3.6, and the H^4 regularity proven in Remark 3.7. □

Bilinear Form Approach

Now, we present an alternative proof of Proposition 3.4 using Lax–Milgram theorem applied to a suitable bilinear form.

Fix $u_0 \in H^1(\mathbb{T}^d)$. Define a bilinear form $B\colon H^2(\mathbb{T}^d) \times H^2(\mathbb{T}^d) \to \mathbb{R}$ and a linear functional $F_{u_0} : H^2(\mathbb{T}^d) \to \mathbb{R}$ by

$$B[u, v] = \int_{\mathbb{T}^d} \epsilon\big(uv + \Delta u \cdot \Delta v\big)\,\mathrm{d}x \quad \text{and} \quad \langle F_{u_0}, v\rangle = \int_{\mathbb{T}^d} F(u_0)v\,\mathrm{d}x$$

for $u, v \in H^2(\mathbb{T}^d)$.

Proposition 3.11 *There exists $\bar{u} \in H^2(\mathbb{T}^d)$ such that*

$$B[\bar{u}, v] = -\langle F_{u_0}, v\rangle \quad \textit{for all } v \in H^2(\mathbb{T}^d). \tag{26}$$

Proof By the Gagliardo–Nirenberg interpolation inequality, we can find a positive constant, C_1, such that the bilinear form B satisfies

$$B[u, u] \geqslant \epsilon C_1 \|u\|^2_{H^2(\mathbb{T}^d)} \quad \text{for all } u \in H^2(\mathbb{T}^d). \tag{27}$$

Moreover, using the Cauchy–Schwarz inequality, there exists another positive constant, C_2, such that

$$|B[u, v]| \leqslant C_2 \|u\|_{H^2(\mathbb{T}^d)} \|v\|_{H^2(\mathbb{T}^d)} \quad \text{for all } u,\ v \in H^2(\mathbb{T}^d). \tag{28}$$

Because $u_0 \in H^1(\mathbb{T}^d)$, $F(u_0)$ is a bounded in L^2. Thus, F_{u_0} is a bounded linear function in $H^2(\mathbb{T}^d)$. Therefore, by the Lax–Milgram theorem, there exists a unique $\bar{u} \in H^2(\mathbb{T}^d)$ satisfying (26). □

Proposition 3.12 *There exists $u \in H^4(\mathbb{T}^d)$ solving*

$$B[u, v] = -\langle F_u, v\rangle \ \textit{for all } v \in H^2(\mathbb{T}^d). \tag{29}$$

Proof To prove the proposition, we apply the fixed-point result in Theorem 3.8. For that, let $\tilde{\mathcal{U}} : H^1(\mathbb{T}^d) \to H^1(\mathbb{T}^d)$ be defined for each $u_0 \in H^1(\mathbb{T}^d)$ by

$$\tilde{\mathcal{U}}[u_0] = \bar{u},$$

where $\bar{u} \in H^2(\mathbb{T}^d)$ is a weak solution to (26).

First, we prove that $\tilde{\mathcal{U}}$ is continuous over $H^1(\mathbb{T}^d)$. Fix $\{u_n\}_{n=1}^{\infty} \subset H^1(\mathbb{T}^d)$ such that $u_n \to u$ in $H^1(\mathbb{T}^d)$ for some $u \in H^1(\mathbb{T}^d)$. Set

$$\bar{u} = \tilde{\mathcal{U}}[u] \ \text{ and } \ \bar{u}_n = \tilde{\mathcal{U}}[u_n].$$

We then have

$$B[\bar{u}, v] = -\langle F_u, v\rangle \ \text{ and } \ B[\bar{u}_n, v] = -\left\langle F_{u_n}, v\right\rangle \ \text{ for all } v \in H^2(\mathbb{T}^d).$$

Subtracting these two equations with $v = \bar{u} - \bar{u}_n \in H^2(\mathbb{T}^d)$, Young's inequality yields

$$\begin{aligned}\epsilon\big(\|\bar{u} - \bar{u}_n\|^2_{L^2(\mathbb{T}^d)} + \|\Delta(\bar{u} - \bar{u}_n)\|^2_{L^2(\mathbb{T}^d)}\big) &\leqslant \int_{\mathbb{T}^d} |F_u - F_{u_n}||\bar{u} - \bar{u}_n|\,\mathrm{d}x \\ &\leqslant \frac{\epsilon}{2}\|\bar{u} - \bar{u}_n\|^2_{L^2(\mathbb{T}^d)} + \frac{C}{\epsilon}\|u - u_n\|^2_{H^1(\mathbb{T}^d)},\end{aligned} \tag{30}$$

from which we obtain that $\|\bar{u} - \bar{u}_n\|^2_{L^2(\mathbb{T}^d)} + \|\Delta(\bar{u} - \bar{u}_n)\|^2_{L^2(\mathbb{T}^d)} \to 0$ as $n \to \infty$. By the Gagliardo–Nirenberg interpolation inequality, it follows that $\|\bar{u} - \bar{u}_n\|_{H^2(\mathbb{T}^d)} \to 0$ as $n \to \infty$. Hence, $\tilde{\mathcal{U}}$ is continuous over $H^1(\mathbb{T}^d)$.

To prove compactness, consider a bounded sequence $\{u_n\}_{n=1}^{\infty} \subset H^1(\mathbb{T}^d)$ and let $\bar{u}_n = \tilde{\mathcal{U}}(u_n)$. Selecting $u = 0$ and $\bar{u} = \tilde{\mathcal{U}}(0)$, the inequality (30) gives the boundedness in $H^2(\mathbb{T}^d)$ of $\bar{u}_n$. Hence, by the Rellich–Kondrachov theorem, $\tilde{\mathcal{U}}$ is compact over $H^1(\mathbb{T}^d)$.

Next, assume that $\lambda \in [0, 1]$ and let $u_\lambda \in H^2(\Omega)$ be such that

$$u_\lambda = \lambda\tilde{\mathcal{U}}[u_\lambda].$$

If $\lambda = 0$, then $u_\lambda = 0$. If $\lambda \in (0, 1]$, we use the fact that $\frac{u_\lambda}{\lambda}$ is a solution to (26) with F_{u_0} replaced by F_{u_λ} to get

$$\int_{\mathbb{T}^d} \epsilon(u_\lambda v + \Delta u_\lambda \Delta v)\,\mathrm{d}x = -\int_{\mathbb{T}^d} \lambda(u_\lambda - b\cdot Du_\lambda - f(x))v\,\mathrm{d}x \ \text{ for all } v \in H^2(\mathbb{T}^d). \tag{31}$$

Setting $v = u_\lambda$ in (31), integrating by parts and using the Gagliardo–Nirenberg interpolation inequality yields

$$\|u_\lambda\|^2_{H^2(\mathbb{T}^d)} \leqslant C_\epsilon,$$

where $C_\epsilon > 0$ is independent of λ.

Applying Theorem 3.8 and the regularity of solutions to (26) proven above, we conclude that there exists $\bar{u} \in H^2(\mathbb{T}^d)$ such that

$$\bar{u} = \tilde{\mathcal{U}}[\bar{u}].$$

Arguing as in Remark 3.7, we see that $\bar{u} \in H^4(\mathbb{T}^d)$. □

Proof of Proposition 3.4 Proposition 3.4 follows directly from the preceding proposition as (29) implies (15). □

Continuity Method

Here, we discuss our last approach to prove Proposition 3.4. We use a continuation argument and the implicit function theorem.

For each $\lambda \in [0, 1]$, we consider the following PDE:

$$\epsilon(u + \Delta^2 u) + (u - \mathbf{b} \cdot Du - \lambda f(x)) = 0, \quad \text{in } \mathbb{T}^d. \tag{32}$$

Note that (20) corresponds to (32) with $\lambda = 1$.

Define

$$S = \Big\{\lambda \in [0, 1] \mid \text{there exists a solution in } H^4(\mathbb{T}^d) \text{ to } (32)\Big\}. \tag{33}$$

We prove next that S is a non-empty set, relatively open and closed in $[0, 1]$; hence, by connectedness, $S = [0, 1]$. In particular, we have a solution, u, to (32) with $\lambda = 1$. This approach to solving PDEs is called the continuity method.

When $\lambda = 0$, $u \equiv 0$ is a solution to (32). Thus, S is nonempty with $0 \in S$. Therefore, it remains to prove closeness and openness, which is done in the following two propositions.

Proposition 3.13 *The set S in (33) is relatively closed in* $[0, 1]$.

Proof Assume that $\{\lambda_n\}_{n=1}^\infty \subset S$ is such that $\lambda_n \to \lambda$ for some $\lambda \in [0, 1]$. We want to show that $\lambda \in S$. Let $\{u_n\}_{n=1}^\infty \subset H^4(\mathbb{T}^d)$ be the corresponding solutions of

$$\epsilon(u_n + \Delta^2 u_n) + (u_n - \mathbf{b} \cdot Du_n - \lambda_n f(x)) = 0, \quad \text{in } \mathbb{T}^d,$$

for all $n \in \mathbb{N}$. Multiplying both sides of the preceding equation by u_n and integrating by parts, we get

$$\epsilon(\|u_n\|^2_{L^2(\mathbb{T}^d)} + \|\Delta u_n\|^2_{L^2(\mathbb{T}^d)}) \leqslant C,$$

where $C > 0$ is independent of n. By the Gagliardo–Nirenberg interpolation inequality, there exists a constant, $C > 0$, such that $\|u_n\|_{H^2(\mathbb{T}^d)} \leqslant C$ for all $n \in \mathbb{N}$. Therefore, extracting a subsequence if necessary, we have $u_n \rightharpoonup u$ in $H^2(\mathbb{T}^d)$ for some $u \in H^2(\mathbb{T}^d)$.

Given $v \in H^2(\mathbb{T}^d)$, u_n satisfies

$$\int_{\mathbb{T}^d} \epsilon(u_n v + \Delta u_n \Delta v)\,dx = \int_{\mathbb{T}^d} (-u_n + \mathbf{b} \cdot Du_n - \lambda_n f(x))v\,dx. \tag{34}$$

Then, taking the limit as $n \to \infty$ in (34), the convergences $u_n \rightharpoonup u$ in $H^2(\mathbb{T}^d)$ and $\lambda_n \to \lambda$ in $\mathbb{R}$ imply that

$$\int_{\mathbb{T}^d} \epsilon(uv + \Delta u \Delta v)\,dx = \int_{\mathbb{T}^d} (-u + \mathbf{b} \cdot Du - \lambda f(x))v\,dx.$$

Because $v \in H^2(\mathbb{T}^d)$ is arbitrary, we conclude that u is a weak solution to (32) in $H^2(\mathbb{T}^d)$. By elliptic regularity, $u \in H^4(\mathbb{T}^d)$. Thus, $\lambda \in S$, which proves that S is closed. □

Proposition 3.14 *The set S in (33) is relatively open in* $[0, 1]$.

Proof Fix $\lambda_0 \in S$, and consider the Fréchet derivative, $\mathcal{L} : H^4(\mathbb{T}^d) \to L^2(\mathbb{T}^d)$, of the map $u \in H^4(\mathbb{T}^d) \mapsto \epsilon(u + \Delta^2 u) + (u - \mathbf{b} \cdot Du - \lambda_0 f) \in L^2(\mathbb{T}^d)$. We have

$$\mathcal{L}(v) = \epsilon(v + \Delta^2 v) + (v - \mathbf{b} \cdot Dv), \quad v \in H^4(\mathbb{T}^d).$$

We aim at proving that $\mathcal{L}$ is an isomorphism from $H^4(\mathbb{T}^d)$ to $L^2(\mathbb{T}^d)$, in which case we conclude that S is relatively open by the implicit function theorem in Banach spaces (see, for example, [22]).

To prove that S is injective, assume that $v_1, v_2 \in H^4(\mathbb{T}^d)$ satisfy $\mathcal{L}(v_1) = \mathcal{L}(v_2)$. Then, multiplying $\mathcal{L}(v_1) - \mathcal{L}(v_2)$ by $v_1 - v_2$, integrating over $\mathbb{T}^d$, an integrating by parts and the condition $\operatorname{div}(\mathbf{b}) = 0$ imply that

$$\|v_1 - v_2\|^2_{L^2(\mathbb{T}^d)} + \epsilon(\|v_1 - v_2\|^2_{L^2(\mathbb{T}^d)} + \|\Delta(v_1 - v_2)\|^2_{L^2(\mathbb{T}^d)}) = 0,$$

from which we get $v_1 - v_2 \equiv 0$. Hence, $\mathcal{L}$ is injective.

Next, we prove that $\mathcal{L}$ is surjective. Fix $w_0 \in L^2(\mathbb{T}^d)$. Using the Lax–Milgram theorem, we prove next that there exists $v_0 \in H^4(\mathbb{T}^d)$ such that $\mathcal{L}(v_0) = w_0$. Define $B : H^2(\mathbb{T}^d) \times H^2(\mathbb{T}^d) \to \mathbb{R}$ by

$$B[v_1, v_2] = \int_{\mathbb{T}^d} \left[\epsilon(v_1 v_2 + \Delta v_1 \Delta v_2) + v_1 v_2 - \mathbf{b} \cdot Dv_1 v_2\right] dx, \quad v_1,\ v_2 \in H^2(\mathbb{T}^d).$$

Arguing as before, B is a bilinear form satisfying

$$B[v, v] = \epsilon(\|v\|_{L^2(\mathbb{T}^d)}^2 + \|\Delta v\|_{L^2(\mathbb{T}^d)}^2) + \|v\|_{L^2(\mathbb{T}^d)}^2 \geqslant C_\epsilon \|v\|_{H^2(\mathbb{T}^d)}$$

for all $v \in H^2(\mathbb{T}^d)$, where $C_\epsilon > 0$ is independent of v. Moreover,

$$|B[v_1, v_2]| \leqslant C\|v_1\|_{H^2(\mathbb{T}^d)}\|v_2\|_{H^2(\mathbb{T}^d)}$$

for all $v_1, v_2 \in H^2(\mathbb{T}^d)$, where $C > 0$ is independent of v_1 and v_2. Therefore, by the Lax–Milgram theorem, there exists $v_0 \in H^2(\mathbb{T}^d)$ such that

$$B[v_0, w] = \int_{\mathbb{T}^d} w_0 \, w \, dx \quad \text{for all } w \in H^2(\mathbb{T}^d).$$

Because $v_0 \in H^2(\mathbb{T}^d)$, we have $g = -v_0 + \mathbf{b} \cdot Dv_0 + w_0 \in L^2(\mathbb{T})$. Then, for all $w \in H^2(\mathbb{T}^d)$, v_0 satisfies

$$\int_{\mathbb{T}^d} \epsilon\big(v_0 w + \Delta v_0 \Delta w\big) \, dx = \int_{\mathbb{T}^d} g \, w \, dx.$$

By the elliptic regularity theory, we have $v_0 \in H^4(\mathbb{T}^d)$, from which we conclude that $\mathcal{L}$ is surjective. □

Since S is nonempty, relatively open, and closed in $[0, 1]$, we have $S = [0, 1]$. Thus, there exists a solution, $u \in H^4(\mathbb{T}^d)$, to (32) with $\lambda = 1$.

Proof of Proposition 3.4 Proposition 3.4 follows directly from the preceding results as (32) implies (15). □

4 Existence of Solutions to Stationary MFG Problems

Here, we use Minty's method to construct weak solutions to stationary MFGs. The main difficulty in applying Minty's method is to build regularizations that preserve monotonicity, have uniform a priori bounds, and for which we can prove the existence of solutions. We show how to use the techniques developed in the prior section to achieve these goals for MFGs.

Our model problem is the MFG problem in (2) with the corresponding operator F given by (3). As usual in the MFG theory, we work under the following general assumptions. The first one concerns the (mild) regularity of the problem data.

Assumption 1 The Hamiltonian, H, is of class C^2 and convex. The coupling, g, is a non-negative continuous increasing function. The source term, ϕ, belongs to $C(\mathbb{T}^d)$ and is such that $\phi > 0$ and $\int_{\mathbb{T}^d} \phi \, dx = 1$.

The previous assumption is satisfied in many MFGs that appear in the literature. For example, $H(p) = \frac{|p|^2}{2}$, $g(m) = m$, and $\phi \equiv 1$ satisfy our requirements.

The convexity of the Hamiltonian reflects the fact that the mean-field game is associated with a control problem. The monotonicity of g encodes that the agents are crowd-averse. The non-negativity assumption on g is used to simplify the presentation. However, our techniques may be adapted to handle, for example, $g(m) = \log m$. The source ϕ represents the incoming flow of agents, which we chose to be normalized to 1.

Assumption 2 There exist constants, $\alpha > 1$ and $C > 0$, such that

$$-H(p) + D_pH(p) \cdot p \geqslant \frac{1}{C}|p|^\alpha - C.$$

By the definition of the Hamiltonian through the Legendre transform in (1), the preceding assumption is a lower bound in the Lagrangian when written in terms of the momentum, p, variable.

Assumption 3 Let α be as in Assumption 2. There exists a constant, $C > 0$, such that

$$H(p) \geqslant \frac{1}{C}|p|^\alpha - C.$$

In classical mechanics, the Hamiltonian is the energy of the system. The preceding bound imposes the growth of the energy as the momentum increases. The last two assumptions are satisfied for $L(v) = \frac{|v|^{\alpha'}}{\alpha'}$ with corresponding Hamiltonian $H(p) = \frac{|p|^\alpha}{\alpha}$, where $\frac{1}{\alpha} + \frac{1}{\alpha'} = 1$.

Assumption 4 For each $\delta \in (0, 1)$, there exists a constant, $C_\delta > 0$, such that, for all $m \in L^1(\mathbb{T}^d)$ with $m \geqslant 0$, we have

$$\int_{\mathbb{T}^d} m \, dx + \int_{\mathbb{T}^d} g(m) \, dx \leqslant \delta \int_{\mathbb{T}^d} mg(m) \, dx + C_\delta.$$

Remark 4.1 In Assumption 4, we assume that g is non-negative. However, we may relax this condition on g and assume instead that g is bounded from below. An example of a function bounded from below satisfying the estimate in Assumption 4 is given by $g(m) = m \ln m$ for $m \in \mathbb{R}_+$. Moreover, though not bounded from below, the function g defined by $g(m) = \log(m)$ satisfies Assumption 4, and also Assumption 5 below, provided that $m > 0$ almost everywhere in $\mathbb{T}^d$.

Assumption 5 Assume that $\{m_j\}_{j=1}^\infty \subset L^1(\Omega)$ satisfies $m_j \geqslant 0$ and

$$\sup_{j \in \mathbb{N}} \int_{\mathbb{T}^d} m_j g(m_j) \, dx < \infty.$$

Then, there is a subsequence, $\{m_{j_i}\}_{i=1}^\infty$, that weakly converges in $L^1(\mathbb{T}^d)$ to some $m \in L^1(\mathbb{T}^d)$.

Observe that F in (3) is well defined in $D(F) = H^{2k}(\mathbb{T}^d; \mathbb{R}_0^+) \times H^{2k}(\mathbb{T}^d)$ taking values in $L^2(\mathbb{T}^d) \times L^2(\mathbb{T}^d)$ for $k \in \mathbb{N}$ large enough. For our purposes, we work with $2k > \frac{d}{2} + 3$. Moreover, we see that F is monotone in $L^2(\mathbb{T}^d) \times L^2(\mathbb{T}^d)$ and (6) holds.

The main goal of this section is to prove existence of weak solutions to (2) in the sense of the following definition.

Definition 4.2 A pair $(m, u) \in L^1(\mathbb{T}^d; \mathbb{R}_0^+) \times L^1(\mathbb{T}^d)$ is a weak solution to (2), that is to

$$F \begin{bmatrix} m \\ u \end{bmatrix} = 0, \tag{35}$$

if

$$\left\langle F \begin{bmatrix} w \\ v \end{bmatrix}, \begin{bmatrix} m \\ u \end{bmatrix} - \begin{bmatrix} w \\ v \end{bmatrix} \right\rangle \leqslant 0 \tag{36}$$

for all $(w, v) \in H^{2k}(\mathbb{T}^d; \mathbb{R}_0^+) \times H^{2k}(\mathbb{T}^d)$.

Remark 4.3 In this paper, we use $\langle \cdot, \cdot \rangle$ to denote the L^2 inner product. By abuse of notation, whenever f and g are functions, not necessarily in L^2 but whose product is in L^1, we write $\langle f, g \rangle = \int_{\mathbb{T}^d} fg \, \mathrm{d}x$. In the inequality (36), we use this convention. In particular, because $F(w, v)$ is continuous whenever $(w, v) \in H^{2k}(\mathbb{T}^d; \mathbb{R}_0^+) \times H^{2k}(\mathbb{T}^d)$, the corresponding integral in (36) is well defined.

In the following theorem, we prove the existence of weak solutions.

Theorem 4.4 *Under Assumptions 1–5, there exists a weak solution* $(m, u) \in L^1(\mathbb{T}^d; \mathbb{R}_0^+) \times W^{1,\alpha}(\mathbb{T}^d)$ *to (2).*

We discuss two approaches to prove this theorem. Both introduce a regularized MFG operator F_ϵ and establish the existence of a solution of

$$F_\epsilon \begin{bmatrix} m_\epsilon \\ u_\epsilon \end{bmatrix} = 0, \tag{37}$$

with $m_\epsilon \geqslant 0$. In both cases, the fundamental difficulty lies in proving the non-negativity of the population density, m_ϵ. The first approach gives the existence of a weak solution of (37) with $m_\epsilon \geqslant 0$ using a variational approach combined with a bilinear form techniques. The proof corresponding to this approach requires the following additional assumption.

Assumption 6 Let α be as in Assumption 2. If $\alpha \geqslant d$, let $\alpha^* = +\infty$. Otherwise, let α^* be the Sobolev conjugate exponent of α,

$$\frac{1}{\alpha^*} = \frac{1}{\alpha} - \frac{1}{d}.$$

There exist $q < \alpha^*$ and $C > 0$ such that, for all $m \in L^1(\mathbb{T}^d)$,

$$\int_{\mathbb{T}^d} m^{q'}\,\mathrm{d}x \leqslant C \int_{\mathbb{T}^d} mg(m)\,\mathrm{d}x + C,$$

where q' is the conjugate exponent of q. Moreover,

$$H(p) \leqslant C(|p|^\alpha + 1). \tag{38}$$

The second approach constructs a smooth solution to (37) with $m_\epsilon > 0$ using the continuity method and does not require the preceding assumption. Hence, Theorem 4.4 is stated under Assumptions 1–5.

Our first choice of F_ϵ is as follows. For $\epsilon \in (0, 1)$, we set

$$F_\epsilon \begin{bmatrix} m \\ u \end{bmatrix} = \begin{bmatrix} -u - H(Du) + g(m) - V + \epsilon(m + \Delta^{2k} m) \\ m - \operatorname{div}\big(mD_pH(Du)\big) - \phi + \epsilon(u + \Delta^{2k} u) \end{bmatrix} \tag{39}$$

for $(m, u) \in H^{4k}(\mathbb{T}^d; \mathbb{R}_0^+) \times H^{4k}(\mathbb{T}^d)$.

A key feature of this regularization is that it preserves monotonicity. In fact, for all $(m_1, u_1), (m_2, u_2) \in H^{4k}(\mathbb{T}^d; \mathbb{R}_0^+) \times H^{4k}(\mathbb{T}^d)$, we have

$$\begin{aligned} &\left\langle F_\epsilon \begin{bmatrix} m_1 \\ u_1 \end{bmatrix} - F_\epsilon \begin{bmatrix} m_2 \\ u_2 \end{bmatrix}, \begin{bmatrix} m_1 \\ u_1 \end{bmatrix} - \begin{bmatrix} m_2 \\ u_2 \end{bmatrix} \right\rangle \\ &= \left\langle F \begin{bmatrix} m_1 \\ u_1 \end{bmatrix} - F \begin{bmatrix} m_2 \\ u_2 \end{bmatrix}, \begin{bmatrix} m_1 \\ u_1 \end{bmatrix} - \begin{bmatrix} m_2 \\ u_2 \end{bmatrix} \right\rangle \\ &\quad + \epsilon \int_{\mathbb{T}^d} \left[(m_1 - m_2)^2 + (\Delta^k m_1 - \Delta^k m_2)^2\right] \mathrm{d}x \geqslant 0, \end{aligned} \tag{40}$$

where we used the monotonicity of F.

Remark 4.5 In general, if $(m, u) \in H^{2k}(\mathbb{T}^d; \mathbb{R}_0^+) \times H^{2k}(\mathbb{T}^d)$, then $F_\epsilon[m, u]$ may not belong to $L^2(\mathbb{T}^d) \times L^2(\mathbb{T}^d)$ due to the presence of the $4kth$-derivatives. However, to simplify the notation, given $(m_1, u_1), (m_2, u_2) \in H^{2k}(\mathbb{T}^d; \mathbb{R}_0^+) \times H^{2k}(\mathbb{T}^d)$, we write

$$\left\langle F_\epsilon \begin{bmatrix} m_1 \\ u_1 \end{bmatrix}, \begin{bmatrix} m_2 \\ u_2 \end{bmatrix} \right\rangle \quad \text{in place of} \quad \left\langle F \begin{bmatrix} m_1 \\ u_1 \end{bmatrix}, \begin{bmatrix} m_2 \\ u_2 \end{bmatrix} \right\rangle + \epsilon \int_{\mathbb{T}^d} \left(m_1 m_2 + \Delta^k m_1 \Delta^k m_2\right) \mathrm{d}x.$$

Note that if $(m_1, u_1) \in H^{4k}(\mathbb{T}^d; \mathbb{R}_0^+) \times H^{4k}(\mathbb{T}^d)$, then an integration by parts shows that these two expressions coincide.

We further observe that, with this convention in mind, (40) holds for all (m_1, u_1), $(m_2, u_2) \in H^{2k}(\mathbb{T}^d; \mathbb{R}_0^+) \times H^{2k}(\mathbb{T}^d)$.

In our second regularization, we address the non-negativity constraint of m by adding a penalty term, p_ϵ, to the first equation. This technique was first used in this context in [26]. More precisely, for each $\epsilon \in (0, 1)$, we fix a non-decreasing C^∞ function, $p_\epsilon \colon (0, \infty) \to (-\infty, 0]$, such that

$$p_\epsilon(t) = \begin{cases} -\frac{1}{t^{d+1}} & \text{if } 0 < t \leqslant \frac{\epsilon}{2} \\ 0 & \text{if } t \geqslant \epsilon. \end{cases} \tag{41}$$

Note that $p_\epsilon(t) \to -\infty$ as $t \to 0^+$. Then, we consider the following alternative choice for F_ϵ:

$$F_\epsilon \begin{bmatrix} m \\ u \end{bmatrix} = \begin{bmatrix} -u - H(Du) + g(m) - V + p_\epsilon(m) + \epsilon(m + \Delta^{2k} m) \\ m - \operatorname{div}\left(m D_p H(Du)\right) - \phi + \epsilon(u + \Delta^{2k} u) \end{bmatrix}, \tag{42}$$

defined for u and m smooth with $m > 0$.

Remark 4.6 As with the previous regularization in (39), the regularization in (42) is also monotone. This can be seen following the same steps used to establish (40).

4.1 Variational Problem and Bilinear Form Approaches

Here, we combine the variational approach with the bilinear form arguments discussed in Sects. 3.1–3.1.

First, we choose $k \in \mathbb{N}$ such that $2k > \frac{d}{2} + 3$, fix $(m_1, u_1) \in H^{2k-1}(\mathbb{T}^d; \mathbb{R}_0^+) \times H^{2k-1}(\mathbb{T}^d)$, and define

$$\begin{aligned} f_1[m_1, u_1] &= -u_1 - H(Du_1) + g(m_1) - V, \\ f_2[m_1, u_1] &= -m_1 + \operatorname{div}(m_1 D_p H(Du_1)) + \phi. \end{aligned}$$

Consider the variational problem of finding $m \in H^{2k}(\mathbb{T}^d; \mathbb{R}_0^+)$ such that

$$J_\epsilon(m) = \inf_{w \in H^{2k}(\mathbb{T}^d; \mathbb{R}_0^+)} J_\epsilon(w),$$

$$\text{where } J_\epsilon(w) = \int_{\mathbb{T}^d} \left(\frac{\epsilon}{2}\left(w^2 + (\Delta^k w)^2\right) + f_1[m_1, u_1] w \right) \mathrm{d}x. \tag{43}$$

Next, we define the bilinear form

$$B_\epsilon[v_1, v_2] = \epsilon \int_{\mathbb{T}^d} (v_1 v_2 + \Delta^k v_1 \Delta^k v_2)\, \mathrm{d}x \quad \text{for } v_1, v_2 \in H^{2k}(\mathbb{T}^d),$$

and consider the problem of finding $u \in H^{2k}(\mathbb{T}^d)$ such that

$$B_\epsilon[u, v] = \langle f_2[m_1, u_1], v \rangle \quad \text{for all } v \in H^{2k}(\mathbb{T}^d), \tag{44}$$

where, as before, $\langle \cdot, \cdot \rangle$ stands for the L^2-inner product. This bilinear form satisfies the conditions in Lax–Milgram's theorem for the Hilbert space $H^{2k}(\mathbb{T}^d)$. Namely, it is coercive and bounded as in (27) and (28).

To prove existence and uniqueness of solutions to these two problems, we first observe that $(m_1, u_1) \in C^1(\mathbb{T}^d) \times C^1(\mathbb{T}^d)$ by Morrey's embedding theorem. Hence,

$f_1[m_1, u_1] \in C(\mathbb{T}^d)$ and $f_2[m_1, u_1] \in L^2(\mathbb{T}^d)$, provided that Assumption 1 holds. Therefore, we can argue as in Sects. 3.1 and 3.1 to show the following proposition.

Proposition 4.7 *Suppose that Assumption 1 holds and that* $(m_1, u_1) \in H^{2k-1}(\mathbb{T}^d) \times H^{2k-1}(\mathbb{T}^d)$. *Then, there exists a unique solution to problems (43) and (44).*

Next, to address the regularized problem (37) with F_ϵ given by (39), we define a map $A : H^{2k-1}(\mathbb{T}^d; \mathbb{R}_0^+) \times H^{2k-1}(\mathbb{T}^d) \to H^{2k-1}(\mathbb{T}^d; \mathbb{R}_0^+) \times H^{2k-1}(\mathbb{T}^d)$ by setting, for each $(m_1, u_1) \in H^{2k-1}(\mathbb{T}^d; \mathbb{R}_0^+) \times H^{2k-1}(\mathbb{T}^d)$,

$$A \begin{bmatrix} m_1 \\ u_1 \end{bmatrix} = \begin{bmatrix} m_2 \\ u_2 \end{bmatrix},$$

where $m_2 \in H^{2k}(\mathbb{T}^d; \mathbb{R}_0^+)$ is the unique solution to (43) and $u_2 \in H^{2k}(\mathbb{T}^d)$ is the unique solution to (44). As in Sect. 3.1, we want to show that A has a fixed point by using Theorem 3.8. Thus, we need to establish the following result.

Proposition 4.8 *Suppose that Assumption 1 holds. Then, the map A is continuous and compact over* $H^{2k-1}(\mathbb{T}^d; \mathbb{R}_0^+) \times H^{2k-1}(\mathbb{T}^d)$.

Proof To prove continuity, we take a convergent sequence (m_n, u_n) in $H^{2k-1}(\mathbb{T}^d; \mathbb{R}_0^+) \times H^{2k-1}(\mathbb{T}^d)$. The maps $f_1(m_n, u_n)$ and $f_2(m_n, u_n)$ converge in $C(\mathbb{T}^d)$ and in $L^2(\mathbb{T}^d)$, respectively. Thus, it suffices to argue as in Sects. 3.1 and 3.1 to establish continuity.

To address compactness, we show next that the solutions to (43) and (44) are automatically in H^{2k} and are bounded in terms of the L^2 norms of f_1 and f_2, respectively. In turn, these norms of f_1 and f_2 are bounded by the norms of m_1 and u_1 in H^{2k-1}. For the solution m_2 of the variational problem (43), we have

$$\int_{\mathbb{T}^d} \left(\frac{\epsilon}{2} (m_2^2 + (\Delta^k m_2)^2) + f_1[m_1, u_1] m_2 \right) \mathrm{d}x \leqslant 0$$

by using $w = 0$ as a competitor in (43). By the Cauchy inequality,

$$\int_{\mathbb{T}^d} f_1[m_1, u_1] m_2 \,\mathrm{d}x \leqslant C_\epsilon \| f_1 \|^2_{L^2(\mathbb{T}^d)} + \frac{\epsilon}{4} \| m_2 \|^2_{L^2(\mathbb{T}^d)}.$$

Combining the two previous inequalities with the Gagliardo–Nirenberg inequality, we obtain $\|m_2\|^2_{H^{2k}(\mathbb{T}^d)} \leqslant C_\epsilon \|f_1\|^2_{L^2(\mathbb{T}^d)}$. For the solution u_2 of (44), we use $v = u_2$ in (44) and argue analogously to get $\|u_2\|^2_{H^{2k}(\mathbb{T}^d)} \leqslant C_\epsilon \|f_2\|^2_{L^2(\mathbb{T}^d)}$. By the Rellich–Kondrachov theorem, we have $H^{2k}(\mathbb{T}^d; \mathbb{R}_0^+) \times H^{2k}(\mathbb{T}^d) \subset\subset H^{2k-1}(\mathbb{T}^d; \mathbb{R}_0^+) \times H^{2k-1}(\mathbb{T}^d)$. This compact embedding combined with the preceding a priori estimates in $H^{2k}(\mathbb{T}^d) \times H^{2k}(\mathbb{T}^d)$ on the solutions allow us to conclude that A is compact over $H^{2k-1}(\mathbb{T}^d; \mathbb{R}_0^+) \times H^{2k-1}(\mathbb{T}^d)$. □

Next, we address the a priori bound needed to apply Schaeffer's fixed point theorem.

Proposition 4.9 *Suppose that Assumptions 1–4 hold. Fix* $\lambda \in [0, 1]$ *and assume that* (m_λ, u_λ) *satisfies*

$$\begin{bmatrix} m_\lambda \\ u_\lambda \end{bmatrix} = \lambda A \begin{bmatrix} m_\lambda \\ u_\lambda \end{bmatrix}. \tag{45}$$

Then, (m_λ, u_λ) *is uniformly bounded in* $H^{2k}(\mathbb{T}^d) \times H^{2k}(\mathbb{T}^d)$ *with respect to* λ.

Proof If $\lambda = 0$, then $(m_\lambda, u_\lambda) \equiv 0$. If $\lambda \in (0, 1]$, then $\frac{m_\lambda}{\lambda}$ solves (43) with $f_1[m_1, u_1]$ replaced by $f_1[m_\lambda, u_\lambda]$. Consequently, for all $w \in H^{2k}(\mathbb{T}^d; \mathbb{R}_0^+)$, we have

$$\int_{\mathbb{T}^d} \Big(\epsilon\big(m_\lambda(w - m_\lambda) + \Delta^k m_\lambda \Delta^k (w - m_\lambda)\big) + \lambda f_1[m_\lambda, u_\lambda](w - m_\lambda)\Big)\, \mathrm{d}x \geqslant 0. \tag{46}$$

Similarly, $\frac{u_\lambda}{\lambda}$ solves (44) with $f_2[m_1, u_1]$ replaced by $f_2[m_\lambda, u_\lambda]$; that is,

$$B_\epsilon[u_\lambda, v] = \lambda \langle f_2[m_\lambda, u_\lambda], v\rangle \quad \text{for all } v \in H^{2k}(\mathbb{T}^d). \tag{47}$$

Taking $w = 0$ in (46) and $v = u_\lambda$ in (47) and combining the resulting estimates with the convexity of H, Assumptions 2, 3, and 4, and the monotonicity of the system, we obtain the bound

$$\begin{aligned}\lambda &\int_{\mathbb{T}^d} \left(m_\lambda g(m_\lambda) + m_\lambda |Du_\lambda|^\alpha + \phi |Du_\lambda|^\alpha\right) \mathrm{d}x \\ &+ \epsilon \int_{\mathbb{T}^d} \left(m_\lambda^2 + (\Delta^k m_\lambda)^2 + u_\lambda^2 + (\Delta^k u_\lambda)^2\right) \mathrm{d}x \leqslant C,\end{aligned} \tag{48}$$

where $C > 0$ is independent of λ and ϵ. Because ϕ and g are non-negative by Assumption 1, the left-hand side of (48) is non-negative. Therefore, the Gagliardo–Nirenberg inequality yields uniform H^{2k} bounds for (m_λ, u_λ) with respect to λ. □

Corollary 4.10 *Suppose that Assumptions 1–4 hold. Then, there exists* $(m_\epsilon, u_\epsilon) \in H^{2k}(\mathbb{T}^d; \mathbb{R}_0^+) \times H^{2k}(\mathbb{T}^d)$ *satisfying*

$$\begin{bmatrix} m_\epsilon \\ u_\epsilon \end{bmatrix} = A \begin{bmatrix} m_\epsilon \\ u_\epsilon \end{bmatrix} \tag{49}$$

and the bound

$$\begin{aligned}&\int_{\mathbb{T}^d} \left(m_\epsilon g(m_\epsilon) + m_\epsilon |Du_\epsilon|^\alpha + \phi |Du_\lambda|^\alpha\right) \mathrm{d}x \\ &+ \epsilon \int_{\mathbb{T}^d} \left(m_\epsilon^2 + (\Delta^k m_\epsilon)^2 + u_\epsilon^2 + (\Delta^k u_\epsilon)^2\right) \mathrm{d}x \leqslant C,\end{aligned} \tag{50}$$

where C *is independent of* ϵ. *In particular,* (m_ϵ, u_ϵ) *is a weak solution to (37).*

Proof By Proposition 4.8, the map A is continuous and compact over $H^{2k-1}(\mathbb{T}^d; \mathbb{R}_0^+) \times H^{2k-1}(\mathbb{T}^d)$. Further, by Proposition 4.9, we have uniform bounds in $H^{2k-1}(\mathbb{T}^d; \mathbb{R}_0^+) \times H^{2k-1}(\mathbb{T}^d)$ for solutions to (45). Consequently, for each $\epsilon \in (0, 1)$, we can apply Theorem 3.8 to obtain a fixed point, (m_ϵ, u_ϵ), with $m_\epsilon \geqslant 0$, satisfying (49). Moreover, by the preceding proposition, $(m_\epsilon, u_\epsilon) \in H^{2k}(\mathbb{T}^d) \times H^{2k}(\mathbb{T}^d)$ and using (48) with $\lambda = 1$ gives (50).

From the above, it follows that (m_ϵ, u_ϵ) is a solution to the variational inequality associated with (37); that is, for all $(w, v) \in H^{2k}(\mathbb{T}^d; \mathbb{R}_0^+) \times H^{2k}(\mathbb{T}^d)$, we have

$$\left\langle F_\epsilon \begin{bmatrix} m_\epsilon \\ u_\epsilon \end{bmatrix}, \begin{bmatrix} m_\epsilon \\ u_\epsilon \end{bmatrix} - \begin{bmatrix} w \\ v \end{bmatrix} \right\rangle \leqslant 0, \tag{51}$$

which concludes the proof. □

Remark 4.11 By choosing $w = 0$ and $w = 2m_\epsilon$ in (46) for $\lambda = 1$, we conclude that

$$\int_{\mathbb{T}^d} \left(\epsilon\left(m_\epsilon^2 + (\Delta^k m_\epsilon\right)^2 + f_1[m_\epsilon, u_\epsilon]m_\epsilon\right) \mathrm{d}x = 0.$$

The preceding identity combined with (50) and Assumptions 3 and 5 yields

$$\int_{\mathbb{T}^d} u_\epsilon m_\epsilon \,\mathrm{d}x \leqslant C \tag{52}$$

for some positive constant C independent of ϵ. Moreover, under condition (38) in Assumption 6, we further obtain that

$$-\int_{\mathbb{T}^d} u_\epsilon m_\epsilon \,\mathrm{d}x \leqslant C, \tag{53}$$

where C does not depend on ϵ.

On the other hand, by choosing $v = 1$ in (47) for $\lambda = 1$, we get

$$\int_{\mathbb{T}^d} m_\epsilon \,\mathrm{d}x = \int_{\mathbb{T}^d} \phi \,\mathrm{d}x - \epsilon \int_{\mathbb{T}^d} u_\epsilon \,\mathrm{d}x.$$

Because of (50), we have that $\epsilon \int_{\mathbb{T}^d} u_\epsilon \,\mathrm{d}x = O(\sqrt{\epsilon})$. Therefore, taking into account that $\int_{\mathbb{T}^d} \phi \,\mathrm{d}x = 1$, we conclude that

$$\lim_{\epsilon \to 0} \int_{\mathbb{T}^d} m_\epsilon \,\mathrm{d}x = 1. \tag{54}$$

Next, we prove Theorem 4.4 under the additional Assumption 6; that is, that there exists a weak solution, $(m, u) \in L^{q'}(\mathbb{T}^d; \mathbb{R}_0^+) \times W^{1,\alpha}(\mathbb{T}^d)$ with $\int_{\mathbb{T}^d} m \,\mathrm{d}x = 1$, to (2) provided that Assumptions 1–6 hold.

Proof of Theorem 4.4 under Assumption 6 Let $(m_\epsilon, u_\epsilon) \in H^{2k}(\mathbb{T}^d) \times H^{2k}(\mathbb{T}^d)$, with $m_\epsilon \geqslant 0$, be the weak solution to (49) given by Corollary 4.10. We first establish some compactness properties of (m_ϵ, u_ϵ).

From (50), Assumption 5, the condition $m_\epsilon \geqslant 0$, and extracting a subsequence if necessary, we can find $m \in L^1(\mathbb{T}^d)$ such that $m \geqslant 0$ and

$$m_\epsilon \rightharpoonup m \text{ in } L^1(\mathbb{T}^d).$$

Moreover, assuming in addition Assumption 6,

$$m_\epsilon \rightharpoonup m \text{ in } L^{q'}(\mathbb{T}^d). \tag{55}$$

By combining Assumption 1 with (50), we obtain further that

$$\int_{\mathbb{T}^d} |Du_\epsilon|^\alpha \, \mathrm{d}x \leqslant C, \tag{56}$$

where $C > 0$ is uniformly bounded with respect to ϵ. Define

$$\tilde{u}_\epsilon = u_\epsilon - \int_{\mathbb{T}^d} u_\epsilon \, \mathrm{d}x.$$

Then, (56) and the Poincaré–Wirtinger inequality yield

$$\|\tilde{u}_\epsilon\|^\alpha_{L^\alpha(\mathbb{T}^d)} \leqslant C \|D\tilde{u}_\epsilon\|^\alpha_{L^\alpha(\mathbb{T}^d)} \leqslant C.$$

Furthermore, recalling Assumption 6, $\tilde{u}_\epsilon$ converges, up to a subsequence, strongly in $L^q(\mathbb{T}^d)$ to some function $\tilde{u}$. Thus, using (55),

$$\lim_{\epsilon \to 0} \int_{\mathbb{T}^d} \tilde{u}_\epsilon m_\epsilon \, \mathrm{d}x = \int_{\mathbb{T}^d} \tilde{u} m \, \mathrm{d}x. \tag{57}$$

Additionally, by (52) and (53), we have that

$$\left| \int_{\mathbb{T}^d} u_\epsilon m_\epsilon \, \mathrm{d}x \right|$$

is bounded uniformly in ϵ. This uniform bound, the identity

$$\int_{\mathbb{T}^d} u_\epsilon m_\epsilon \, \mathrm{d}x = \left(\int_{\mathbb{T}^d} u_\epsilon \, \mathrm{d}x \right) \left(\int_{\mathbb{T}^d} m_\epsilon \, \mathrm{d}x \right) + \int_{\mathbb{T}^d} \tilde{u}_\epsilon m_\epsilon \, \mathrm{d}x,$$

(54), and (57) imply that $\int_{\mathbb{T}^d} u_\epsilon$ converges, up to a subsequence. Therefore, u_ϵ converges to some function u in $L^q(\mathbb{T}^d)$ and weakly in $W^{1,\alpha}(\mathbb{T}^d)$.

Next, given $(w, v) \in H^{2k}(\mathbb{T}^d; \mathbb{R}^+_0) \times H^{2k}(\mathbb{T}^d)$, we observe that (51) and the monotonicity of F_ϵ (see Remark 4.5) yield

$$\begin{aligned} 0 &\geqslant \left\langle F_\epsilon \begin{bmatrix} m_\epsilon \\ u_\epsilon \end{bmatrix}, \begin{bmatrix} m_\epsilon \\ u_\epsilon \end{bmatrix} - \begin{bmatrix} w \\ v \end{bmatrix} \right\rangle \geqslant \left\langle F_\epsilon \begin{bmatrix} w \\ v \end{bmatrix}, \begin{bmatrix} m_\epsilon \\ u_\epsilon \end{bmatrix} - \begin{bmatrix} w \\ v \end{bmatrix} \right\rangle \\ &= \left\langle F \begin{bmatrix} w \\ v \end{bmatrix}, \begin{bmatrix} m_\epsilon \\ u_\epsilon \end{bmatrix} - \begin{bmatrix} w \\ v \end{bmatrix} \right\rangle + h_\epsilon, \end{aligned} \tag{58}$$

where

$$h_\epsilon = \epsilon \int_{\mathbb{T}^d} \Big(w(m_\epsilon - w) + \Delta^k w \Delta^k (m_\epsilon - w) \\ + v(u_\epsilon - v) + \Delta^k v \Delta^k (u_\epsilon - v) \Delta^k (u_\epsilon - v) \Big) \, \mathrm{d}x.$$

From (50), we have uniform L^2-bounds for $(\sqrt{\epsilon} m_\epsilon, \sqrt{\epsilon}\Delta^k m_\epsilon, \sqrt{\epsilon} u_\epsilon, \sqrt{\epsilon}\Delta^k u_\epsilon)$; thus, using Hölder's inequality, we have

$$\lim_{\epsilon \to 0} h_\epsilon = 0.$$

Consequently, taking the limit in (58) as $\epsilon \to 0$, we conclude that

$$0 \geqslant \left\langle F \begin{bmatrix} w \\ v \end{bmatrix}, \begin{bmatrix} m \\ \tilde{u} \end{bmatrix} - \begin{bmatrix} w \\ v \end{bmatrix} \right\rangle$$

for all $(w, v) \in H^{2k}(\mathbb{T}^d; \mathbb{R}_0^+) \times H^{2k}(\mathbb{T}^d)$. Thus, $(m, u) \in L^{q'}(\mathbb{T}^d) \times W^{1,\alpha}(\mathbb{T}^d)$ is a weak solution to (2) with $\int_{\mathbb{T}^d} m \, \mathrm{d}x = 1$ by (54) and (55). □

Remark 4.12 Without assuming Assumption 6, we use the arguments in the preceding proof to show the existence of a weak solution in a slightly different sense, under the assumptions of Theorem 4.4. Precisely, there exists $(m, \tilde{u}) \in L^1(\mathbb{T}^d; \mathbb{R}_0^+) \times W^{1,\alpha}(\mathbb{T}^d)$ such that (36) holds for all $(w, v) \in \big\{ w \in H^{2k}(\mathbb{T}^d; \mathbb{R}_0^+) | \int_{\mathbb{T}^d} w \, \mathrm{d}x = 1 \big\} \times H^{2k}(\mathbb{T}^d)$.

Indeed, recalling that $f_2[w, v] = -w + \operatorname{div}(w D_p H(Dv)) - \phi$, we have

$$\begin{aligned} \langle -f_2[w, v], u_\epsilon \rangle &= \langle -f_2[w, v], \tilde{u}_\epsilon \rangle + \int_{\mathbb{T}^d} u_\epsilon \, \mathrm{d}x \int_{\mathbb{T}^d} (w - \phi) \, \mathrm{d}x \\ &= \langle -f_2[w, v], \tilde{u}_\epsilon \rangle . \end{aligned}$$

Consequently,

$$\left\langle F \begin{bmatrix} w \\ v \end{bmatrix}, \begin{bmatrix} m_\epsilon \\ u_\epsilon \end{bmatrix} - \begin{bmatrix} w \\ v \end{bmatrix} \right\rangle = \left\langle F \begin{bmatrix} w \\ v \end{bmatrix}, \begin{bmatrix} m_\epsilon \\ \tilde{u}_\epsilon \end{bmatrix} - \begin{bmatrix} w \\ v \end{bmatrix} \right\rangle.$$

It suffices then to pass to the limit in the preceding expression.

4.2 Continuity Method

Here, we use the continuity method to prove Theorem 4.4. For that, we introduce the following λ-dependent regularized MFG:

$$F_\epsilon^\lambda \begin{bmatrix} m \\ u \end{bmatrix} = \begin{bmatrix} 0 \\ 0 \end{bmatrix}, \tag{59}$$

where $\lambda \in [0, 1]$ and, for $(m, u) \in C^{4k}(\mathbb{T}^d) \times C^{4k}(\mathbb{T}^d)$ with $m > 0$,

$$F_\epsilon^\lambda \begin{bmatrix} m \\ u \end{bmatrix} = \begin{bmatrix} -u - \lambda H(Du) + \lambda g(m) - \lambda V + p_\epsilon(m) + \epsilon(m + \Delta^{2k} m) \\ m - \lambda \operatorname{div}\left(m D_p H(Du)\right) - \lambda\phi - (1-\lambda) + \epsilon(u + \Delta^{2k} u) \end{bmatrix}. \tag{60}$$

We want to show that there exists a solution to the preceding problem when $\lambda = 1$, which corresponds to the regularized MFG (37) with F_ϵ given by (42). To this end, we set

$$\Lambda = \big\{\lambda \in [0,1] \;|\text{problem (4.25) admits a smooth a solution,} \\ (m_\lambda, u_\lambda) \in C^\infty(\mathbb{T}^d) \times C^\infty(\mathbb{T}^d), \text{ with } \min_{\mathbb{T}^d} m_\lambda > 0\big\}, \tag{61}$$

and start by proving that Λ is a nonempty set.

Fix $\epsilon \in (0,1)$ and let $\lambda = 0$. For $c \in [0, \frac{1}{\epsilon})$, define

$$i(c) = -c + p_\epsilon(1 - \epsilon c) + \epsilon(1 - \epsilon c).$$

We have $i \in C([0, \frac{1}{\epsilon}))$, $i(0) = \epsilon > 0$, and $\lim_{c \to (\frac{1}{\epsilon})^-} i(c) = -\infty$. Hence, there is $c^* \in [0, \frac{1}{\epsilon})$ such that $i(c^*) = 0$. Therefore, $(m, u) = (1 - \epsilon c^*, c^*)$ is a solution to (59) with $\lambda = 0$; note further that $m = 1 - \epsilon c^* > 0$. Hence,

$$\Lambda \neq \emptyset. \tag{62}$$

Next, our goal is to prove that Λ is both relatively open and closed, from which we deduce that $\Lambda = [0,1]$. In particular, there exists a solution for $\lambda = 1$ as stated in the next proposition.

Proposition 4.13 *Under Assumptions 1–4, there exists a smooth solution to (37) with F_ϵ given by (42).*

Proof As explained before, (37) corresponds to (59) with $\lambda = 1$. So the result follows from Λ being non-empty by (62) and both relatively open and closed. The closedeness is proved in Lemma 4.14, while the openess is proven in Lemma 4.18. □

Λ Is Closed

The main result in this subsection is the following lemma.

Lemma 4.14 *Under Assumptions 1–4, the set Λ introduced in (61) is closed.*

The proof of Lemma 4.14 relies on the establishing some a priori uniform bounds on the solutions of (59) as stated in the two following lemmas.

Proposition 4.15 *Let $\epsilon \in (0,1)$ and $\lambda \in [0,1]$, and assume that (m_λ, u_λ), with $m_\lambda > 0$, is a classical solution to (59). Then, under Assumptions 1–4, there exists a constant, $C > 0$, which is independent of ϵ and λ, such that (m_λ, u_λ) satisfies*

$$\int_{\mathbb{T}^d} -p_\epsilon(m_\lambda)(\lambda\phi + 1 - \lambda)\,\mathrm{d}x + \int_{\mathbb{T}^d} \big(p_\epsilon(m_\lambda)(m_\lambda - \epsilon) + \lambda g(m_\lambda)m_\lambda\big)\,\mathrm{d}x$$
$$+ \int_{\mathbb{T}^d} \big(\lambda m_\lambda |Du_\lambda|^\alpha + \lambda(\lambda\phi + 1 - \lambda)|Du_\lambda|^\alpha\big)\,\mathrm{d}x \tag{63}$$
$$+ \epsilon \int_{\mathbb{T}^d} (m_\lambda^2 + (\Delta^k m_\lambda)^2 + u_\lambda^2 + (\Delta^k u_\lambda)^2) \leqslant C.$$

Moreover, (m_λ, u_λ) *is bounded in* $H^{2k}(\mathbb{T}^d) \times H^{2k}(\mathbb{T}^d)$*, uniformly with respect to* λ*.*

Proof Fix $\epsilon \in (0, 1)$ and $\lambda \in [0, 1]$, and assume that there exists $(m_\lambda, u_\lambda) \in C^\infty(\mathbb{T}^d) \times C^\infty(\mathbb{T}^d)$ solving (59) and such that $m_\lambda > 0$.

We start multiplying the first equation in (59) by $(m_\lambda - \epsilon - \lambda\phi - 1 + \lambda)$ and the second one by u_λ; we then integrate over $\mathbb{T}^d$ and add the resulting identities to obtain

$$\int_{\mathbb{T}^d} \Big[\lambda m_\lambda\big(-H(Du_\lambda) + D_pH(Du_\lambda)\cdot Du_\lambda\big) + \lambda(\epsilon + \lambda\phi + 1 - \lambda)H(Du_\lambda)\Big]\,\mathrm{d}x$$
$$+ \int_{\mathbb{T}^d} \Big[\lambda m_\lambda g(m_\lambda) - (\lambda\phi + 1 - \lambda)p_\epsilon(m_\lambda) + (m_\lambda - \epsilon)p_\epsilon(m_\lambda)\Big]\,\mathrm{d}x$$
$$+ \epsilon \int_{\mathbb{T}^d} \Big[m_\lambda^2 + (\Delta^k m_\lambda)^2 + u_\lambda^2 + (\Delta^k u_\lambda)^2\Big]$$
$$= \int_{\mathbb{T}^d} \Big[\lambda(\epsilon + \lambda\phi + 1 - \lambda)g(m_\lambda) + \lambda m_\lambda V(x) - \lambda(\epsilon + \lambda\phi + 1 - \lambda)V(x)\Big]\,\mathrm{d}x$$
$$+ \int_{\mathbb{T}^d} \Big[\epsilon(\epsilon + \lambda\phi + 1 - \lambda)m_\lambda + \epsilon\lambda\Delta^k m_\lambda \Delta^k \phi\Big]\,\mathrm{d}x - \epsilon \int_{\mathbb{T}^d} u_\lambda\,\mathrm{d}x. \tag{64}$$

Next, we observe that $0 \leqslant \epsilon + \lambda\phi + 1 - \lambda \leqslant 2 + \|\phi\|_\infty$ because $\epsilon \in (0, 1)$, $\phi \geqslant 0$, and $\lambda \in [0, 1]$. Thus, using Assumptions 1–3 first and then Assumption 4 with $\delta = \frac{1}{4}$, we can estimate the first integral on the left-hand side of (64) as follows:

$$\int_{\mathbb{T}^d} \Big[\lambda m_\lambda\big(-H(Du_\lambda) + D_pH(Du_\lambda)\cdot Du_\lambda\big) + \lambda(\epsilon + \lambda\phi + 1 - \lambda)H(Du_\lambda)\Big]\,\mathrm{d}x$$
$$\geqslant \frac{1}{C}\int_{\mathbb{T}^d} \Big[\lambda m_\lambda |Du_\lambda|^\alpha + \lambda(\epsilon + \lambda\phi + 1 - \lambda)|Du_\lambda|^\alpha\Big]\,\mathrm{d}x$$
$$- C\lambda \int_{\mathbb{T}^d} m_\lambda\,\mathrm{d}x + \lambda(\epsilon + \lambda\phi + 1 - \lambda)C$$
$$\geqslant \frac{1}{C}\int_{\mathbb{T}^d} \Big[\lambda m_\lambda |Du_\lambda|^\alpha + \lambda(\epsilon + \lambda\phi + 1 - \lambda)|Du_\lambda|^\alpha\Big]\,\mathrm{d}x - \frac{\lambda}{4}\int_{\mathbb{T}^d} m_\lambda g(m_\lambda)\,\mathrm{d}x - \widetilde{C}, \tag{65}$$

where $\widetilde{C}$ is a positive constant that is independent of ϵ and of λ. Similarly, the right-hand side of (64) can be estimated as follows:

$$\begin{aligned}
&\int_{\mathbb{T}^d}\Big[\lambda(\epsilon+\lambda\phi+1-\lambda)g(m_\lambda)+\lambda m_\lambda V(x)-\lambda(\epsilon+\lambda\phi+1-\lambda)V(x)\Big]\,\mathrm{d}x\\
&\quad+\int_{\mathbb{T}^d}\Big[\epsilon(\epsilon+\lambda\phi+1-\lambda)m_\lambda+\epsilon\lambda\Delta^k m_\lambda\Delta^k\phi\Big]\,\mathrm{d}x-\epsilon\int_{\mathbb{T}^d}u_\lambda\,\mathrm{d}x \qquad (66)\\
&\quad\leqslant\frac{\lambda}{4}\int_{\mathbb{T}^d}m_\lambda g(m_\lambda)\,\mathrm{d}x+\frac{\epsilon}{2}\int_{\mathbb{T}^d}\Big(m_\lambda^2+(\Delta^k m_\lambda)^2+u_\lambda^2\Big)\,\mathrm{d}x+\widehat{C},
\end{aligned}$$

where $\widetilde{C}$ is a positive constant that is independent of ϵ and of λ. Then, it follows from (64)–(66) that

$$\begin{aligned}
&\frac{1}{C}\int_{\mathbb{T}^d}\Big[\lambda m_\lambda|Du_\lambda|^\alpha+\lambda(\epsilon+\lambda\phi+1-\lambda)|Du_\lambda|^\alpha\Big]\,\mathrm{d}x\\
&\quad+\int_{\mathbb{T}^d}\Big[\frac{\lambda}{2}m_\lambda g(m_\lambda)-(\lambda\phi+1-\lambda)p_\epsilon(m_\lambda)+(m_\lambda-\epsilon)p_\epsilon(m_\lambda)\Big]\,\mathrm{d}x\\
&\quad+\frac{\epsilon}{2}\int_{\mathbb{T}^d}\Big[m_\lambda^2+(\Delta^k m_\lambda)^2+u_\lambda^2+(\Delta^k u_\lambda)^2\Big]\leqslant\widetilde{C}+\widehat{C}.
\end{aligned}$$

By the properties of p_ϵ, we have that $-p_\epsilon(m_\lambda)\geqslant 0$ and $(m_\lambda-\epsilon)p_\epsilon(m_\lambda)\geqslant 0$. Hence,

$$\epsilon\int_{\mathbb{T}^d}\big[m_\lambda^2+(\Delta^k m_\lambda)^2+u_\lambda^2+(\Delta^k u_\lambda)^2\big]\leqslant\overline{C}$$

for some positive and λ-independent constant $\overline{C}$. The Gagliardo–Nirenberg interpolation inequality then yields bounds in $H^{2k}(\mathbb{T}^d)\times H^{2k}(\mathbb{T}^d)$ for (m_λ,u_λ) that are uniform with respect to λ. □

Remark 4.16 By not using the bound by below for H in Assumption 3 in the estimate (65), we obtain when $\lambda=1$ that

$$\int_{\mathbb{T}^d}H(Du)\,\mathrm{d}x\leqslant C,$$

where the constant does not depend on u or ϵ.

Next, we prove a uniform bound on m_λ from below.

Lemma 4.17 *Let Λ be given by (61) and assume that Assumptions 1–4 hold. Then, there exists a constant, $c>0$, such that*

$$\inf_{\lambda\in\Lambda}\inf_{x\in\mathbb{T}^d}m_\lambda(x)\geqslant c>0.$$

Proof By contradiction, assume that $\inf_{\lambda\in\Lambda}\inf_{\mathbb{T}^d}m_\lambda=0$.

Let $\{\lambda_n\}_{n=1}^\infty\subset\Lambda$ be such that $c_n=\inf_{\mathbb{T}^d}m_{\lambda_n}\to 0$ as $n\to\infty$. By the definition of Λ, there exists $x_n\in\mathbb{T}^d$ such that $c_n=\min_{\mathbb{T}^d}m_{\lambda_n}=m_{\lambda_n}(x_n)>0$. Without loss of generality, we may further assume that $\lambda_n\to\lambda$ for some $\lambda\in[0,1]$. We have from Proposition 4.15 and Morrey's embedding theorem that

$$R = \sup_{n\in\mathbb{N}} \|m_{\lambda_n}\|_{W^{1,\infty}(\mathbb{T}^d)} < \infty.$$

Thus, we can find $n_0 \in \mathbb{N}$ such that $0 < c_n \leqslant \min\{\frac{\epsilon}{4}, \frac{\epsilon}{2\sqrt{d}R}\}$ for all $n \geqslant n_0$. Hence, denoting by Q_n the cube in $\mathbb{R}^d$ with side length c_n and centered at x_n, we have, for all $x \in Q_n$ with $n \geqslant n_0$, that

$$0 < m_{\lambda_n}(x) \leqslant m_{\lambda_n}(x_n) + R|x - x_n| \leqslant c_n + \frac{\sqrt{d}c_n}{2}R \leqslant \frac{\epsilon}{2}.$$

In particular, recalling (41), $-p_\epsilon(m_{\lambda_n}(x)) = \frac{1}{(m_{\lambda_n}(x))^{d+1}}$ for all $x \in Q_n$ with $n \geqslant n_0$.

Then, using the fact that all integral terms on the left-hand side of (63) are non-negative, there exists a constant, $C > 0$, independent of n and for which we have, for all $n \geqslant n_0$, that

$$\begin{aligned} C &\geqslant \int_{\mathbb{T}^d} -p_\epsilon(m_{\lambda_n})(\lambda_n\phi + 1 - \lambda_n)\,\mathrm{d}x \geqslant \int_{Q_n} \frac{1}{m_{\lambda_n}^{d+1}}(\lambda_n\phi + 1 - \lambda_n)\,\mathrm{d}x \\ &\geqslant \frac{c_n^d}{\left(\frac{\sqrt{d}c_n}{2}R + c_n\right)^{d+1}}\left(\lambda_n\phi_{min} + 1 - \lambda_n\right) = \left(\frac{2}{\sqrt{d}R + 2}\right)^{d+1}\frac{1}{c_n}\left(\lambda_n\phi_{min} + 1 - \lambda_n\right), \end{aligned}$$

where $\phi_{min} = \min_{x\in\mathbb{T}^d}\phi(x) > 0$ by Assumption 1. Finally, we observe that $\lim_{n\to\infty}\frac{1}{c_n} = +\infty$ and $\lim_{n\to\infty}(\lambda_n\phi_{min} + 1 - \lambda_n) = \lambda\phi_{min} + 1 - \lambda > 0$ because $\lambda \in [0, 1]$ and $\phi_{min} > 0$. Thus, taking the limit as $n \to \infty$ in the preceding estimate leads to a contradiction, which concludes the proof of the lemma. □

Proof of Lemma 4.14 Let $\{\lambda_n\}_{n=1}^\infty \subset \Lambda$ be such that $\lambda_n \to \lambda$ as $n \to \infty$. Then, $\lambda \in [0, 1]$. To show that $\lambda \in \Lambda$, we use the definition of Λ and Lemma 4.17 to find, for each $n \in \mathbb{N}$, a pair $(m_n, u_n) \in C^\infty(\mathbb{T}^d) \times C^\infty(\mathbb{T}^d)$ with $\min_{\mathbb{T}^d} m_n \geqslant c > 0$ solving (59) with λ replaced by λ_n, where the constant c does not depend on n.

From Proposition 4.15, there exists a subsequence of $\{(m_n, u_n)\}_{n=1}^\infty$, $\{(m_{n_j}, u_{n_j})\}_{j=1}^\infty$, and $(m, u) \in H^{2k}(\mathbb{T}^d) \times H^{2k}(\mathbb{T}^d)$ such that $(m_{n_j}, u_{n_j}) \rightharpoonup (m, u)$ in $H^{2k}(\mathbb{T}^d) \times H^{2k}(\mathbb{T}^d)$. Using the Rellich–Kondrachov theorem and Morrey's theorem, $(m_{n_j}, u_{n_j}) \to (m, u)$ in $C^{2,l}(\mathbb{T}^d) \times C^{2,l}(\mathbb{T}^d)$ for some $l \in (0, 1)$. Thus, we have $\min_{\mathbb{T}^d} m > 0$.

On the other hand, for all $(w, v) \in H^{2k}(\mathbb{T}^d) \times H^{2k}(\mathbb{T}^d)$, we obtain from (59) that

$$\begin{aligned} &\int_{\mathbb{T}^d} \Big(\epsilon(m_{n_j}w + \Delta^k m_{n_j}\Delta^k w) + g_1[m_{n_j}, u_{n_j}]w\Big)\,\mathrm{d}x = 0, \\ &\int_{\mathbb{T}^d} \Big(\epsilon(u_{n_j}v + \Delta^k u_{n_j}\Delta^k v) + g_2[m_{n_j}, u_{n_j}]v\Big)\,\mathrm{d}x = 0, \end{aligned}$$

where $g_1[m_{n_j}, u_{n_j}] = -u_{n_j} - \lambda_{n_j}H(Du_{n_j}) + \lambda_{n_j}g(m_{n_j}) - \lambda_{n_j}V(x) + p_\epsilon(m_{n_j})$ and $g_2[m_{n_j}, u_{n_j}] = m_{n_j} - \lambda_{n_j}\operatorname{div}(m_{n_j}D_pH(Du_{n_j})) - \lambda_{n_j}\phi - (1 - \lambda_{n_j})$. Using the weak convergence in $H^{2k}(\mathbb{T}^d) \times H^{2k}(\mathbb{T}^d)$ and the strong convergence in

$C^{2,l}(\mathbb{T}^d) \times C^{2,l}(\mathbb{T}^d)$ of $\{(m_{n_j}, u_{n_j})\}_{j=1}^{\infty}$ to (m, u) justified above, and the convergence $\lambda_{n_j} \to \lambda \in [0, 1]$, taking a limit as $j \to \infty$ in the previous identities yields

$$\int_{\mathbb{T}^d} \left(\epsilon(mw + \Delta^k m \Delta^k w) + g_1[m, u]w\right) \mathrm{d}x = 0,$$

$$\int_{\mathbb{T}^d} \left(\epsilon(uv + \Delta^k u \Delta^k v) + g_2[m, u]v\right) \mathrm{d}x = 0.$$

Because $g_1[m, u]$, $g_2[m, u] \in C^{0,l}(\mathbb{T}^d)$ for some $l \in (0, 1)$, we have $(m, u) \in C^{4k,l}(\mathbb{T}^d) \times C^{4k,l}(\mathbb{T}^d)$ by standard elliptic-regularity arguments, and (m, u) solves (59). Finally, using a bootstrap argument, we conclude that $(m, u) \in C^{\infty}(\mathbb{T}^d) \times C^{\infty}(\mathbb{T}^d)$. Hence, $\lambda \in \Lambda$, which proves that Λ is closed. □

Λ Is Open

Next, we show that Λ is relatively open.

Lemma 4.18 *Under Assumptions 1–4, the set Λ introduced in (61) is relatively open in* $[0, 1]$.

Proof Fix $\lambda_0 \in \Lambda$ and let (m_0, u_0) be a smooth solution to (59) with $\lambda = \lambda_0$ such that $\inf_{\mathbb{T}^d} m_0 > 0$. We will find a neighborhood of λ_0 in $[0, 1]$ contained in Λ using the implicit function theorem in Banach spaces (see, for example, [22]). Here, we use the implicit function theorem in $C^{4k}(\mathbb{T}^d) \times C^{4k}(\mathbb{T}^d)$. That is, we regard the map F as a map from $C^{4k}(\mathbb{T}^d) \times C^{4k}(\mathbb{T}^d)$ to $C(\mathbb{T}^d) \times C(\mathbb{T}^d)$.

First, recalling (60), we compute the Fréchet derivative, $\mathcal{L}$, of the mapping $(m, u) \mapsto F_{\epsilon}^{\lambda_0}(m, u)$ at (m_0, u_0). For each $(w, v) \in C^{4k}(\mathbb{T}^d) \times C^{4k}(\mathbb{T}^d)$, we have

$$\mathcal{L}\begin{bmatrix} w \\ v \end{bmatrix} = \begin{bmatrix} -v - \lambda_0 D_p H(Du_0) \cdot Dv + \lambda_0 g'(m_0)w + p_{\epsilon}'(m_0)w + \epsilon(w + \Delta^{2k} w) \\ w - \lambda_0 \operatorname{div}\left(w D_p H(Du_0) + m_0 D_{pp}^2 H(Du_0) Dv\right) + \epsilon(v + \Delta^{2k} v) \end{bmatrix}.$$

We want to prove that $\mathcal{L}$ is an isomorphism from $C^{4k}(\mathbb{T}^d) \times C^{4k}(\mathbb{T}^d)$ to $C(\mathbb{T}^d) \times C(\mathbb{T}^d)$.

Because $\mathcal{L}$ is linear, to show that $\mathcal{L}$ is injective is equivalent to showing that its kernel is trivial. For that, we assume that $(w, v) \in C^{4k}(\mathbb{T}^d) \times C^{4k}(\mathbb{T}^d)$ satisfies $\mathcal{L}[w, v] = 0$. Recalling that H is convex, g and p_{ϵ} are increasing, and $m_0 > 0$, it follows that

$$\lambda_0 m_0 (Dv)^T D_{pp}^2 H(Du_0) Dv \geqslant 0, \quad \lambda_0 g'(m_0) w^2 \geqslant 0, \quad p_{\epsilon}'(m_0) w^2 \geqslant 0. \tag{67}$$

Therefore, taking the inner product between $\mathcal{L}[w, v]$ and (w, v), an integration by parts yields

$$
\begin{aligned}
0 &= \int_{\mathbb{T}^d} \Big(\lambda_0\big(g'(m_0)w^2 + m_0(Dv)^T D^2_{pp} H(Du_0)Dv\big) + p'_\epsilon(m_0)w^2\Big)\, \mathrm{d}x \\
&\quad + \epsilon \int_{\mathbb{T}^d} \Big(w^2 + (\Delta^k w)^2 + v^2 + (\Delta^k v)^2\Big)\, \mathrm{d}x \\
&\geqslant \epsilon \int_{\mathbb{T}^d} \Big(w^2 + (\Delta^k w)^2 + v^2 + (\Delta^k v)^2\Big)\, \mathrm{d}x \geqslant 0,
\end{aligned}
$$

from which we obtain that $(w, v) \equiv (0, 0)$.

Next, we show that $\mathcal{L}$ is surjective. For this, we need to show that given $(h_1, h_2) \in C(\mathbb{T}^d) \times C(\mathbb{T}^d)$ there exists $(\bar{w}, \bar{v}) \in C^{4k}(\mathbb{T}^d) \times C^{4k}(\mathbb{T}^d)$ that solves $\mathcal{L}(\bar{w}, \bar{v}) = (h_1, h_2)$. As a first step, we a find a solution to this problem in $H^{2k}(\mathbb{T}^d) \times H^{2k}(\mathbb{T}^d)$.

For $(w_1, v_1), (w_2, v_2) \in H^{2k}(\mathbb{T}^d) \times H^{2k}(\mathbb{T}^d)$, define

$$
\begin{aligned}
B\left[\left[\begin{matrix} w_1 \\ v_1 \end{matrix}\right], \left[\begin{matrix} w_2 \\ v_2 \end{matrix}\right]\right] = \int_{\mathbb{T}^d} \Big(& - v_1 w_2 - \lambda_0 D_p H(Du_0) \cdot Dv_1 w_2 \\
& + \lambda_0 g'(m_0) w_1 w_2 + p'_\epsilon(m_0) w_1 w_2 \\
& + w_1 v_2 - \lambda_0 \operatorname{div}\big(w_1 D_p H(Du_0) + m_0 D^2_{pp} H(Du_0) Dv_1\big) v_2\Big)\, \mathrm{d}x \\
& + \epsilon \int_{\mathbb{T}^d} (w_1 w_2 + \Delta^k w_1 \Delta^k w_2 + v_1 v_2 + \Delta^k v_1 \Delta^k v_2)\, \mathrm{d}x.
\end{aligned}
$$

We consider the problem of finding $(\bar{w}, \bar{v}) \in H^{2k}(\mathbb{T}^d) \times H^{2k}(\mathbb{T}^d)$ such that, for all $(\xi, \eta) \in H^{2k}(\mathbb{T}^d) \times H^{2k}(\mathbb{T}^d)$, we have

$$
B\left[\left[\begin{matrix} \bar{w} \\ \bar{v} \end{matrix}\right], \left[\begin{matrix} \xi \\ \eta \end{matrix}\right]\right] = \left\langle \left[\begin{matrix} h_1 \\ h_2 \end{matrix}\right], \left[\begin{matrix} \xi \\ \eta \end{matrix}\right] \right\rangle. \tag{68}
$$

Given $(w, v) \in H^{2k}(\mathbb{T}^d) \times H^{2k}(\mathbb{T}^d)$, an integration by parts and (67) yield

$$
B\left[\left[\begin{matrix} w \\ v \end{matrix}\right], \left[\begin{matrix} w \\ v \end{matrix}\right]\right] \geqslant \epsilon(\|w\|^2_{L^2(\mathbb{T}^d)} + \|v\|^2_{L^2(\mathbb{T}^d)} + \|\Delta^k w\|^2_{L^2(\mathbb{T}^d)} + \|\Delta^k v\|^2_{L^2(\mathbb{T}^d)}).
$$

Thus, by the Gagliardo–Nirenberg interpolation inequality, there exists a constant, $C_\epsilon > 0$, independent of (w, v), for which

$$
B\left[\left[\begin{matrix} w \\ v \end{matrix}\right], \left[\begin{matrix} w \\ v \end{matrix}\right]\right] \geqslant C_\epsilon\big(\|w\|^2_{H^{2k}(\mathbb{T}^d)} + \|v\|^2_{H^{2k}(\mathbb{T}^d)}\big).
$$

On the other hand, the smoothness of (m_0, u_0) and the data, together with the fact that $\min_{\mathbb{T}^d} m_0 > 0$, yields a constant, $C > 0$, such that for all $(w_1, v_1), (w_2, v_2) \in H^{2k}(\mathbb{T}^d) \times H^{2k}(\mathbb{T}^d)$, we have

$$
\left| B\left[\left[\begin{matrix} w_1 \\ v_1 \end{matrix}\right], \left[\begin{matrix} w_2 \\ v_2 \end{matrix}\right]\right] \right| \leqslant C(\|w_1\|_{H^{2k}(\mathbb{T}^d)} + \|v_1\|_{H^{2k}(\mathbb{T}^d)})(\|w_2\|_{H^{2k}(\mathbb{T}^d)} + \|v_2\|_{H^{2k}(\mathbb{T}^d)}).
$$

Hence, applying the Lax–Milgram theorem, there exists $(\bar{w}, \bar{v}) \in H^{2k}(\mathbb{T}^d) \times H^{2k}(\mathbb{T}^d)$ satisfying (68). Finally, using elliptic-regularity and bootstrap arguments, we deduce that $(\bar{w}, \bar{v}) \in C^{4k}(\mathbb{T}^d) \times C^{4k}(\mathbb{T}^d)$, which, together with (68), shows that $\mathcal{L}$ is surjective.

Therefore, by the implicit function theorem, Λ is relatively open in $[0, 1]$. □

Proof of Theorem 4.4

Now, we prove the existence of weak solutions.

Proof of Theorem 4.4 Fix $\epsilon \in (0, 1)$ and let $(m_\epsilon, u_\epsilon) \in C^\infty(\mathbb{T}^d) \times C^\infty(\mathbb{T}^d)$ be the solution of (37) (for F_ϵ given by (42)) constructed in Proposition 4.13. Recall further that $\min_{\mathbb{T}^d} m_\epsilon > 0$ by Lemma (4.17). Next, we establish some uniform estimates in ϵ for (m_ϵ, u_ϵ).

By (63) with $\lambda = 1$, and recalling that $\min_{\mathbb{T}^d} \phi > 0$ by Assumption 1 and $p_\epsilon \leqslant 0$ by construction, we have that

$$\begin{aligned}&\int_{\mathbb{T}^d} (m_\epsilon g(m_\epsilon) + m_\epsilon |Du_\epsilon|^\alpha + |Du_\epsilon|^\alpha - p_\epsilon(m_\epsilon))\,\mathrm{d}x \\ &\quad + \epsilon \int_{\mathbb{T}^d} \left(m_\epsilon^2 + (\Delta^k m_\epsilon)^2 + u_\epsilon^2 + (\Delta^k u_\epsilon)^2\right) \mathrm{d}x \leqslant C,\end{aligned} \tag{69}$$

where C is independent of ϵ and (m_ϵ, u_ϵ). On the other hand, integrating the first equation in (37) over $\mathbb{T}^d$, we get from the conditions $g \geqslant 0$ and $p_\epsilon \leqslant 0$ that

$$\left|\int_{\mathbb{T}^d} u_\epsilon\,\mathrm{d}x\right| \leqslant \left|\int_{\mathbb{T}^d} H(Du_\epsilon)\,\mathrm{d}x\right| + \int_{\mathbb{T}^d} \left(g(m_\epsilon) + |V(x)| - p_\epsilon(m_\epsilon) + \epsilon m_\epsilon\right) \mathrm{d}x.$$

Then, using Remark 4.16, Assumptions 3 and 4, and (69), we conclude that

$$\left|\int_{\mathbb{T}^d} u_\epsilon\,\mathrm{d}x\right| \leqslant C, \tag{70}$$

where $C > 0$ is independent of ϵ and (m_ϵ, u_ϵ). The Poincaré–Wirtinger inequality and (69) and (70) yield

$$\|u_\epsilon\|_{L^\alpha(\mathbb{T}^d)} \leqslant \left\|u_\epsilon - \int_{\mathbb{T}^d} u_\epsilon\,\mathrm{d}x\right\|_{L^\alpha(\mathbb{T}^d)} + C \leqslant C\|Du_\epsilon\|_{L^\alpha(\mathbb{T}^d)} + C \leqslant C.$$

The preceding estimate and (69) show that u_ϵ is bounded in $W^{1,\alpha}(\mathbb{T}^d)$, uniformly with respect to ϵ.

Using the monoticity of F_ϵ addressed in Remark 4.6 and the definition of F in (3), for all $(w, v) \in H^{2k}(\mathbb{T}^d; \mathbb{R}^+) \times H^{2k}(\mathbb{T}^d)$, we have

$$0 = \left\langle F_\epsilon \begin{bmatrix} m_\epsilon \\ u_\epsilon \end{bmatrix}, \begin{bmatrix} m_\epsilon \\ u_\epsilon \end{bmatrix} - \begin{bmatrix} w \\ v \end{bmatrix} \right\rangle \geqslant \left\langle F_\epsilon \begin{bmatrix} w \\ v \end{bmatrix}, \begin{bmatrix} m_\epsilon \\ u_\epsilon \end{bmatrix} - \begin{bmatrix} w \\ v \end{bmatrix} \right\rangle$$

$$= \left\langle F \begin{bmatrix} w \\ v \end{bmatrix}, \begin{bmatrix} m_\epsilon \\ u_\epsilon \end{bmatrix} - \begin{bmatrix} w \\ v \end{bmatrix} \right\rangle + c_\epsilon, \tag{71}$$

where

$$c_\epsilon = \int_{\mathbb{T}^d} \epsilon\big(w(m_\epsilon - w) + \Delta^k w \Delta^k (m - w) + v(u_\epsilon - v) + \Delta^k v \Delta^k (u_\epsilon - v)\big) + p_\epsilon(w)(m_\epsilon - w)\Big)\,\mathrm{d}x.$$

Next, we observe that, extracting a subsequence if necessary, the uniform estimates in ϵ of $\|u_\epsilon\|_{W^{1,\alpha}(\mathbb{T}^d)}$ proved above and Assumption 5 combined with (69) yield

$$\begin{aligned} &m_\epsilon \rightharpoonup m \text{ in } L^1(\mathbb{T}^d),\\ &u_\epsilon \rightharpoonup u \text{ in } W^{1,\alpha}(\mathbb{T}^d),\\ &\sqrt{\epsilon}(\sqrt{\epsilon} m_\epsilon),\ \sqrt{\epsilon}(\sqrt{\epsilon} u_\epsilon) \to 0 \text{ in } H^{2k}(\mathbb{T}^d), \end{aligned}$$

for some $m \in L^1(\mathbb{T}^d)$, with $m \geqslant 0$, and $u \in W^{1,\alpha}(\mathbb{T}^d)$. Because $p_\epsilon(w) \to 0$ in $C(\mathbb{T}^d)$ whenever $w \in H^{2k}(\mathbb{T}^d; \mathbb{R}^+)$, we conclude that

$$\lim_{\epsilon \to 0} c_\epsilon = 0$$

for each $(w, v) \in H^{2k}(\mathbb{T}^d; \mathbb{R}^+) \times H^{2k}(\mathbb{T}^d)$. Therefore, arguing as in Sect. 4.1, we can let $\epsilon \to 0$ in (71) to conclude that

$$0 \geqslant \lim_{\epsilon \to 0} \left(\left\langle F \begin{bmatrix} w \\ v \end{bmatrix}, \begin{bmatrix} m_\epsilon \\ u_\epsilon \end{bmatrix} - \begin{bmatrix} w \\ v \end{bmatrix} \right\rangle + c_\epsilon \right) = \left\langle F \begin{bmatrix} w \\ v \end{bmatrix}, \begin{bmatrix} m \\ u \end{bmatrix} - \begin{bmatrix} w \\ v \end{bmatrix} \right\rangle.$$

Finally, we observe that Assumption 1 allows us to show the non-positivity of the right-hand side of the preceding estimate for all $(w, v) \in H^{2k}(\mathbb{T}^d; \mathbb{R}_0^+) \times H^{2k}(\mathbb{T}^d)$. Thus, $(m, u) \in L^1(\mathbb{T}^d) \times W^{1,\alpha}(\mathbb{T}^d)$ is a weak solution to (35). □

5 Existence of Solutions for Time-Dependent MFGs

In this last section, we explain briefly how the prior techniques based on monotonicity can be used to study the time-dependent MFG in (8) and (9). Many of the steps are similar to the stationary case, so, here, we highlight only the main points and refer the interested reader to [28] for a detailed analysis.

Set $\Omega_T = (0, T) \times \mathbb{T}^d$ and fix $m_0,\ u_T \in C^{4k}(\mathbb{T}^d)$ such that $\min_{x \in \mathbb{T}^d} m_0(x) > 0$ and $\int_{\mathbb{T}^d} m_0(x)\,\mathrm{d}x = 1$, where $k \in \mathbb{N}$ is such that $2k \geqslant \frac{d+1}{2} + 4$. Consider the following (elliptic) regularized version of the MFG in (8):

$$\begin{cases} u_t + \Delta u - H(Du) + g(m) + V + \epsilon\big(m + \sum_{|\beta|=2k} \partial_{t,x}^{2\beta} m\big) = 0 & \text{in } \Omega_T, \\ m_t - \Delta m - \operatorname{div}(m D_p H(Du)) + \epsilon\big(u + \sum_{|\beta|=2k} \partial_{t,x}^{2\beta} u\big) = 0 & \text{in } \Omega_T, \\ m(0,\cdot) = m_0,\ u(T,\cdot) = u_T & \text{in } \mathbb{T}^d, \end{cases} \tag{72}$$

where

$$\partial_{t,x}^{\beta} = \frac{\partial^{|\beta|}}{\partial t^{\beta_0} \partial x_1^{\beta_1} ... \partial x_d^{\beta_d}}, \quad \beta = (\beta_0, \beta_1, ..., \beta_d) \in \mathbb{N}_0^{d+1}, \quad |\beta| = \sum_{i=0}^{d} \beta_i.$$

The presence of high-order derivatives in time requires the following additional boundary conditions. For each $i \in \mathbb{N}$ such that $2 \leqslant i \leqslant 2k$,

$$\begin{cases} \sum_{j=1}^{2k} \partial_t^{2j-1}(M_j m) = 0 & \text{on } \{T\} \times \mathbb{T}^d, \\ \sum_{j=1}^{2k} \partial_t^{2j-1}(M_j u) = 0 & \text{on } \{0\} \times \mathbb{T}^d, \\ \sum_{j=i}^{2k} \partial_t^{2j-1}(M_j m) = 0 & \text{on } \{0, T\} \times \mathbb{T}^d, \\ \sum_{j=i}^{2k} \partial_t^{2j-1}(M_j u) = 0 & \text{on } \{0, T\} \times \mathbb{T}^d, \end{cases}$$

where $M_j = \sum_{|\gamma|=2k-j} \partial_x^{2\gamma}$. These boundary conditions are crucial both to preserve the monotonicity of the original problem and enable a priori estimates for classical and weak solutions of (72).

We show how to address (72) using similar techniques to those in Sect. 4.1 under Assumptions 1–5 (with the obvious modifications to incorporate the dependence in t).

Set

$$\begin{aligned} \mathcal{A} &= \left\{ m \in H^{2k}(\Omega_T) \mid m(0,x) = m_0(x),\ m \geqslant 0 \right\}, \\ \mathcal{B} &= \left\{ u \in H^{2k}(\Omega_T) \mid u(T,x) = u_T \right\}. \end{aligned} \tag{73}$$

We prove the existence of a pair $(m, u) \in \mathcal{A} \times \mathcal{B}$ that is a solution to the variational inequality associated with (72). In particular, this pair satisfies

$$\begin{aligned} &\int_0^T \int_{\mathbb{T}^d} \big(u_t + \Delta u - H(Du) + g(m) + V\big)(w - m)\, \mathrm{d}x\mathrm{d}t \\ &\quad + \epsilon \int_0^T \int_{\mathbb{T}^d} \left[m(w - m) + \sum_{|\beta|=2k} \partial_{t,x}^{\beta} m \partial_{t,x}^{\beta}(w - m) \right] \mathrm{d}x\mathrm{d}t \geqslant 0, \end{aligned} \tag{74}$$

$$\begin{aligned} &\int_0^T \int_{\mathbb{T}^d} \big(m_t - \Delta m - \operatorname{div}\big(m D_p H(x, Du)\big)\big(v - u\big)\, \mathrm{d}x\mathrm{d}t \\ &\quad + \epsilon \int_0^T \int_{\mathbb{T}^d} \left[u(v - u) + \sum_{|\beta|=2k} \partial_{t,x}^{\beta} u \partial_{t,x}^{\beta}(v - u) \right] \mathrm{d}x\mathrm{d}t = 0 \end{aligned} \tag{75}$$

for all $(w, v) \in \mathcal{A} \times \mathcal{B}$.

5.1 A Priori Estimates

To prove a similar estimate to that in (48), choose $w = m_0 \in \mathcal{A}$ in (74) and $v = u_T \in \mathcal{B}$ in (75). Then, adding the resulting inequalities, integrating by parts, and using Assumption 4, we conclude that

$$\begin{aligned}&\int_0^T\int_{\mathbb{T}^d} \left(mg(m) + m|Du|^\alpha + m_0|Du|^\alpha\right) \mathrm{d}x\mathrm{d}t\\&\quad + \epsilon \int_0^T\int_{\mathbb{T}^d} \Big(m^2 + \sum_{|\beta|=2k} (\partial_{t,x}^\beta m)^2 + u^2 + \sum_{|\beta|=2k} (\partial_{t,x}^\beta u)^2\Big) \mathrm{d}x\mathrm{d}t\\&\leqslant C\big(\|m\|_{L^1(\Omega_T)} + \|Du\|_{L^1(\Omega_T)} + 1\big).\end{aligned}$$

By Assumption 4 and the fact that $\min_{x\in\mathbb{T}^d} m_0(x) \geqslant c > 0$ for some $c > 0$, we can find for each $\delta \in (0, 1)$ a constant, C_δ, independent of ϵ and (m, u), for which we have

$$\int_0^T\int_{\mathbb{T}^d} (m + |Du|)\,\mathrm{d}x\mathrm{d}t \leqslant \delta\Big(\int_0^T\int_{\mathbb{T}^d} (mg(m) + |Du|^\alpha)\,\mathrm{d}x\mathrm{d}t\Big) + C_\delta.$$

Therefore, choosing $\delta \in (0, 1)$ small enough, there is a constant, $C > 0$, that is independent of ϵ and (m, u), such that

$$\begin{aligned}&\int_0^T\int_{\mathbb{T}^d} (mg(m) + |Du|^\alpha)\,\mathrm{d}x\mathrm{d}t\\&+\epsilon \int_0^T\int_{\mathbb{T}^d} \Big[m^2 + \sum_{|\beta|=2k} (\partial_{t,x}^\beta m)^2 + u^2 + \sum_{|\beta|=2k} (\partial_{t,x}^\beta u)^2\Big] \mathrm{d}x\mathrm{d}t \leqslant C. \qquad (76)\end{aligned}$$

5.2 Variational Problem and Bilinear Form Approach

Let

$$\begin{aligned}\tilde{\mathcal{A}} &= \left\{\tilde m \in H^{2k}(\Omega_T) | \tilde m(0, x) = 0,\ \tilde m + m_0 \geqslant 0\right\},\\ \tilde{\mathcal{B}} &= \left\{\tilde u \in H^{2k}(\Omega_T) | \tilde u(T, x) = 0\right\}.\end{aligned}$$

For $(m, u) \in \mathcal{A} \times \mathcal{B}$, we write $u = u_T + \tilde u$ and $m = m_0 + \tilde m$ with $(\tilde m, \tilde u) \in \tilde{\mathcal{A}} \times \tilde{\mathcal{B}}$. Accordingly, we modify H, g, and V, as follows:

$$\tilde H(D\tilde u) = H(Du_T + D\tilde u),\quad \tilde g(\tilde m) = g(m_0 + \tilde m),\quad \tilde V(t, x) = V(t, x) + \Delta u_T.$$

Set

$$\begin{aligned}f_1[\tilde m, \tilde u] &= \tilde u_t + \Delta \tilde u - \tilde H(D\tilde u) + \tilde g(\tilde m) + \tilde V,\\ f_2[\tilde m, \tilde u] &= \tilde m_t - \Delta \tilde m - \Delta m_0 - \operatorname{div}((m_0 + \tilde m) D_p \tilde H(D\tilde u)).\end{aligned}$$

Fix $(\tilde{m}_1, \tilde{u}_1) \in H^{2k-1}(\Omega_T) \times H^{2k-1}(\Omega_T)$, and for $w, v_1, v_2 \in H^{2k}(\Omega_T)$, define

$$J(w) = \int_0^T \int_{\mathbb{T}^d} \left[\frac{\epsilon}{2} \left(w^2 + \sum_{|\beta|=2k} (\partial_{t,x}^\beta w)^2 \right) + f_1[\tilde{m}_1, \tilde{u}_1] w \right] \mathrm{d}x\mathrm{d}t,$$

$$B[v_1, v_2] = \epsilon \int_0^T \int_{\mathbb{T}^d} \left[v_1 v_2 + \sum_{|\beta|=2k} \partial_{t,x}^\beta v_1 \partial_{t,x}^\beta v_2 \right] \mathrm{d}x\mathrm{d}t.$$

As in Sect. 4.1, we search for a pair, $(\tilde{m}, \tilde{u}) \in \tilde{\mathcal{A}} \times \tilde{\mathcal{B}}$, satisfying

$$J(\tilde{m}) = \inf_{w \in \tilde{\mathcal{A}}} J(w) \tag{77}$$

and

$$B[\tilde{u}, v] = \langle f_2[\tilde{m}_1, \tilde{u}_1], v \rangle \quad \text{for all } v \in \tilde{\mathcal{B}}. \tag{78}$$

It can be checked that similar arguments to those of Sect. 4.1 allow us to prove that such a pair $(\tilde{m}, \tilde{u}) \in \tilde{\mathcal{A}} \times \tilde{\mathcal{B}}$ exists, being $\tilde{m}$ the unique minimizer to (77) and $\tilde{u}$ the unique weak solution to (78), respectively.

5.3 Existence of a Fixed-Point

Next, we construct a solution, $(\tilde{m}, \tilde{u})$, to (74) and (75). As in the previous section, we use Theorem 3.8. Let $\widehat{\mathcal{A}}$ and $\widehat{\mathcal{B}}$ be given by

$$\widehat{\mathcal{A}} = \left\{ \tilde{m} \in H^{2k-1}(\Omega_T) \mid \tilde{m}(0, x) = 0,\ \tilde{m} + m_0 \geqslant 0 \right\},$$
$$\widehat{\mathcal{B}} = \left\{ \tilde{u} \in H^{2k-1}(\Omega_T) \mid \tilde{u}(T, x) = 0 \right\}.$$

Define $A : \widehat{\mathcal{A}} \times \widehat{\mathcal{B}} \to \widehat{\mathcal{A}} \times \widehat{\mathcal{B}}$ by

$$\begin{bmatrix} \tilde{m}_2 \\ \tilde{u}_2 \end{bmatrix} = A \begin{bmatrix} \tilde{m}_1 \\ \tilde{u}_1 \end{bmatrix},$$

where $(\tilde{m}_2, \tilde{u}_2)$ satisfies (77) and (78), respectively. Using a similar argument to the one in Sect. 4.1 combined with the a priori estimates (76), we have that A is continuous and compact.

Next, let

$$S = \left\{ (\tilde{m}, \tilde{u}) \in \widehat{\mathcal{A}} \times \widehat{\mathcal{B}} \,\middle|\, \begin{bmatrix} \tilde{m} \\ \tilde{u} \end{bmatrix} = \lambda A \begin{bmatrix} \tilde{m} \\ \tilde{u} \end{bmatrix} \text{ for some } \lambda \in [0, 1] \right\}.$$

We want to prove that S is bounded. When $\lambda = 0$, we have $(\tilde{m}, \tilde{u}) \equiv (0, 0)$. When $\lambda \in (0, 1]$, we proceed as follows. Assume that there exists $(\tilde{m}_\lambda, \tilde{u}_\lambda) \in S$. Because $(\frac{\tilde{m}_\lambda}{\lambda}, \frac{\tilde{u}_\lambda}{\lambda})$ is a pair of the minimizer to (77) and the weak solution to (78), we have, for all $w \in \mathcal{A}_0$ and $v \in \mathcal{B}_0$, that

$$\int_0^T\int_{\mathbb{T}^d}\Big[\epsilon\Big(\tilde{m}_\lambda(\lambda w - \tilde{m}_\lambda) + \sum_{|\beta|=2k}\partial^\beta_{t,x}\tilde{m}_\lambda\partial^\beta_{t,x}(\lambda w - \tilde{m}_\lambda)\Big)$$
$$+\lambda f_1[\tilde{m}_\lambda, \tilde{u}_\lambda](\lambda w - \tilde{m}_\lambda)\Big]\,\mathrm{d}x\mathrm{d}t \geqslant 0,$$
$$\int_0^T\int_{\mathbb{T}^d}\Big[\tilde{u}_\lambda v + \sum_{|\beta|=2k}\partial^\beta_{t,x}\tilde{u}_\lambda\partial^\beta_{t,x}v\Big]\,\mathrm{d}x\mathrm{d}t = \lambda\int_0^T\int_{\mathbb{T}^d} f_2[\tilde{m}_\lambda, \tilde{u}_\lambda]v\,\mathrm{d}x\mathrm{d}t.$$

Choosing $w = 0$ and $v = \tilde{u}_\lambda$ in the preceding inequalities, and adding the resulting estimates, we get

$$\lambda\int_0^T\int_{\mathbb{T}^d}(m_\lambda g(m_\lambda) + (m_\lambda + m_0)|Du_\lambda|^\alpha)\,\mathrm{d}x\mathrm{d}t$$
$$+\epsilon\int_0^T\int_{\mathbb{T}^d}\Big[\tilde{m}_\lambda^2 + \sum_{|\beta|=2k}(\partial^\beta_{t,x}\tilde{m}_\lambda)^2 + \tilde{u}_\lambda^2 + \sum_{|\beta|=2k}(\partial^\beta_{t,x}\tilde{u}_\lambda)^2\Big]\,\mathrm{d}x\mathrm{d}t \leqslant C,$$

where C is independent of λ. By the Gagliardo–Nirenberg interpolation inequality, $(\tilde{m}_\lambda, \tilde{u}_\lambda)$ is bounded in $H^{2k}(\Omega_T) \times H^{2k}(\Omega_T)$, uniformly with respect to λ. Therefore, S is bounded in $\widehat{\mathcal{A}} \times \widehat{\mathcal{B}}$. Hence, we can apply Theorem 3.8 to A and, thus, we construct a fixed point, $(\tilde{m}^*, \tilde{u}^*) \in \widetilde{\mathcal{A}} \times \widetilde{\mathcal{B}}$. Therefore $(m^*, u^*) = (\tilde{m}^*, \tilde{u}^*) + (m_0, u_T)$ solves (74) and (75).

5.4 Weak Solutions

Let (m_ϵ, u_ϵ) solve (74) and (75). We first establish some estimates for (m_ϵ, u_ϵ) that are uniform in ϵ.

By Assumption 5 and (76), there is $m \in L^1(\Omega_T)$ such that, extracting a subsequence if necessary, $m_\epsilon \rightharpoonup m$ in $L^1(\Omega_T)$. For $t \in (0, T)$, define

$$\langle u_\epsilon\rangle(t) = \int_{\mathbb{T}^d} u_\epsilon(t, x)\,\mathrm{d}x.$$

Setting $\overline{u}_\epsilon = u_\epsilon - \langle u_\epsilon\rangle$ and combining the Poincaré–Wirtinger inequality with (76), we obtain

$$\|\overline{u}_\epsilon\|^\alpha_{L^\alpha(\Omega_T)} \leqslant C\|D\overline{u}_\epsilon\|^\alpha_{L^\alpha(\Omega_T)} \leqslant C$$

for some constant, $C > 0$, uniform with respect to ϵ. Thus, there is $\overline{u} \in L^{\alpha}([0,T]; W^{1,\alpha}(\mathbb{T}^d))$ such that, up to a subsequence, $\overline{u}_\epsilon \rightharpoonup \overline{u}$ in $L^{\alpha}([0,T]; W^{1,\alpha}(\mathbb{T}^d))$. In particular, using (76) once more, we have

$$\begin{aligned} m_\epsilon \rightharpoonup m \text{ in } L^1(\Omega_T),\ \overline{u}_\epsilon \rightharpoonup \overline{u} \text{ in } W^{1,\alpha}(\Omega_T),\ \epsilon m_\epsilon \to 0 \text{ in } H^{2k}(\Omega_T), \\ \text{and } \epsilon u_\epsilon \to 0 \text{ in } H^{2k}(\Omega_T). \end{aligned} \tag{79}$$

Let $\mathcal{A}^*$ be the subset of $\mathcal{A}$ (see (73)) given by

$$\mathcal{A}^* = \left\{ w \in \mathcal{A} \;\middle|\; \int_{\mathbb{T}^d} w \,\mathrm{d}x = 1 \text{ for a.e. } t \in [0,T] \right\}.$$

As before, for $\epsilon \in (0,1)$ and (m,u) sufficiently smooth, we define two operators, F and F_ϵ, associated with the original time-dependent MFG and the regularized one, respectively, by setting

$$F\begin{bmatrix} m \\ u \end{bmatrix} = \begin{bmatrix} u_t + \Delta u - H(Du) + g(m) + V \\ m_t - \Delta m - \operatorname{div}\big(m D_p H(Du)\big) \end{bmatrix},$$
$$F_\epsilon\begin{bmatrix} m \\ u \end{bmatrix} = \begin{bmatrix} u_t + \Delta u - H(Du) + g(m) + V + \epsilon(m + \Delta^{2k} m) \\ m_t - \Delta m - \operatorname{div}\big(m D_p H(Du)\big) + \epsilon(u + \Delta^{2k} u) \end{bmatrix}.$$

Under Assumptions 1–4 (adapted to the time-dependent case), and having in mind Remark 4.5, it can be checked that F and F_ϵ are monotone over $\mathcal{A}^* \times \mathcal{B}$. Because (m_ϵ, u_ϵ) satisfies (74) and (75), for all $(w,v) \in \mathcal{A} \times \mathcal{B}$, we have

$$\left\langle F_\epsilon \begin{bmatrix} m_\epsilon \\ u_\epsilon \end{bmatrix}, \begin{bmatrix} m_\epsilon \\ u_\epsilon \end{bmatrix} - \begin{bmatrix} w \\ v \end{bmatrix} \right\rangle \leqslant 0.$$

Fix $(w,v) \in \mathcal{A}^* \times \mathcal{B}$. Then, it can be checked that (see Lemma 7.5 in [28])

$$\left\langle F_\epsilon \begin{bmatrix} w \\ v \end{bmatrix}, \begin{bmatrix} m_\epsilon \\ u_\epsilon \end{bmatrix} - \begin{bmatrix} w \\ v \end{bmatrix} \right\rangle = \left\langle F_\epsilon \begin{bmatrix} w \\ v \end{bmatrix}, \begin{bmatrix} m_\epsilon \\ \overline{u}_\epsilon \end{bmatrix} - \begin{bmatrix} w \\ v \end{bmatrix} \right\rangle,$$

where, we recall, $\overline{u}_\epsilon = u_\epsilon - \langle u_\epsilon \rangle$. Hence, as in (58), for $(w,v) \in \mathcal{A}^* \times \mathcal{B}$, we have that

$$\begin{aligned} 0 &\geqslant \left\langle F_\epsilon \begin{bmatrix} m_\epsilon \\ u_\epsilon \end{bmatrix}, \begin{bmatrix} m_\epsilon \\ u_\epsilon \end{bmatrix} - \begin{bmatrix} w \\ v \end{bmatrix} \right\rangle \geqslant \left\langle F_\epsilon \begin{bmatrix} w \\ v \end{bmatrix}, \begin{bmatrix} m_\epsilon \\ u_\epsilon \end{bmatrix} - \begin{bmatrix} w \\ v \end{bmatrix} \right\rangle \\ &= \left\langle F_\epsilon \begin{bmatrix} w \\ v \end{bmatrix}, \begin{bmatrix} m_\epsilon \\ \overline{u}_\epsilon \end{bmatrix} - \begin{bmatrix} w \\ v \end{bmatrix} \right\rangle \\ &= \left\langle F \begin{bmatrix} w \\ v \end{bmatrix}, \begin{bmatrix} m_\epsilon \\ \overline{u}_\epsilon \end{bmatrix} - \begin{bmatrix} w \\ v \end{bmatrix} \right\rangle + h_\epsilon, \end{aligned} \tag{80}$$

where

$$h_\epsilon = \epsilon \int_0^T \int_{\mathbb{T}^d} \Big[w(w - m_\epsilon) + \sum_{|\beta|=2k} \partial_{t,x}^\beta w \partial_{t,x}^\beta (w - m_\epsilon) + v(v - \overline{u}_\epsilon) + \sum_{|\beta|=2k} \partial_{t,x}^\beta v \partial_{t,x}^\beta (v - \overline{u}_\epsilon) \Big] \mathrm{d}x\mathrm{d}t.$$

By (79) and (80), $\lim_{\epsilon\to 0} h_\epsilon = 0$ and

$$0 \geqslant \lim_{\epsilon\to 0} \left(\left\langle F \begin{bmatrix} w \\ v \end{bmatrix}, \begin{bmatrix} m_\epsilon \\ \overline{u}_\epsilon \end{bmatrix} - \begin{bmatrix} w \\ v \end{bmatrix} \right\rangle + h_\epsilon \right) = \left\langle F \begin{bmatrix} w \\ v \end{bmatrix}, \begin{bmatrix} m \\ \overline{u} \end{bmatrix} - \begin{bmatrix} w \\ v \end{bmatrix} \right\rangle,$$

which is a weak solution to (8) in the spirit of Remark 4.12. To get a weak solution like in Theorem 4.4, where the test functions belong to $\mathcal{A} \times \mathcal{B}$, we would have to prove the convergence of u_ϵ. The authors leave this matter as an open problem for the interested reader.

References

1. Achdou, Y., P. Cardaliaguet, F. Delarue, A. Porretta, and F. Santambrogio. 2020. *Mean field games*, vol. 2281. Lecture notes in mathematics. Cham: Springer; Centro Internazionale Matematico Estivo (C.I.M.E.), Florence, [2020], ed. by Cardaliaguet, Pierre, and Porretta, Alessio. Fondazione CIME/CIME foundation subseries.
2. Almulla, N., R. Ferreira, and D. Gomes. 2017. Two numerical approaches to stationary mean-field games. *Dynamic Games and Applications* 7 (4): 657–682.
3. Ambrose, David M. 2016. Small strong solutions for time-dependent mean field games with local coupling. *Comptes Rendus Mathématique Académie des Sciences Paris* 354 (6): 589–594.
4. Bakaryan, Tigran, Rita Ferreira, and Diogo Gomes. 2022. A potential approach for planning mean-field games in one dimension. *Communications on Pure and Applied Analysis* 21 (6): 2147–2187.
5. Benamou, Jean-David., and Guillaume Carlier. 2015. Augmented Lagrangian methods for transport optimization, mean field games and degenerate elliptic equations. *Journal of Optimization Theory and Applications* 167 (1): 1–26.
6. Bensoussan, A., J. Frehse, and P. Yam. 2013. *Mean field games and mean field type control theory*. Springer briefs in mathematics. New York: Springer.
7. Boccardo, L., L. Orsina, and A. Porretta. 2016. Strongly coupled elliptic equations related to mean-field games systems. *Journal of Differential Equations* 261 (3): 1796–1834.
8. Cannarsa, P., and R. Capuani. 2018. Existence and uniqueness for mean field games with state constraints. In *PDE models for multi-agent phenomena*, vol. 28. Springer INdAM series, 49–71. Cham: Springer.
9. Cannarsa, P., R. Capuani, and P. Cardaliaguet. 2019. $C^{1,1}$-smoothness of constrained solutions in the calculus of variations with application to mean field games. *Mathematics in Engineering* 1 (1): 174–203.
10. Cannarsa, P., R. Capuani, and P. Cardaliaguet. 2021. Mean field games with state constraints: from mild to pointwise solutions of the PDE system. *Calculus of Variations and Partial Differential Equations* 60 (3): Paper No. 108, 33.
11. Cannarsa, P., and C. Mendico. 2020. Mild and weak solutions of mean field game problems for linear control systems. *Minimax Theory and its Applications* 5 (2): 221–250.
12. Cardaliaguet, P. 2011. Notes on mean-field games.
13. Cardaliaguet, P. 2015. *Weak solutions for first order mean field games with local coupling*, vol. 11. Springer INdAM series, 111–158. Cham: Springer.

14. Cardaliaguet, P., P. Garber, A. Porretta, and D. Tonon. 2015. Second order mean field games with degenerate diffusion and local coupling. *NoDEA Nonlinear Differential Equations Applications* 22 (5): 1287–1317.
15. Cardaliaguet, P., and P. J. Graber. 2015. Mean field games systems of first order. *ESAIM Control Optimisation and Calculus of Variations* 21 (3): 690–722.
16. Carmona, R., and F. Delarue. 2018. *Probabilistic theory of mean field games with applications I–II*. Springer.
17. Cirant, M. 2016. Stationary focusing mean-field games. *Communications in Partial Differential Equations* 41 (8): 1324–1346.
18. Cirant, M., R. Gianni, and P. Mannucci. 2020. Short-time existence for a general backward-forward parabolic system arising from mean-field games. *Dynamic Games and Applications* 10 (1): 100–119.
19. Cirant, M., and A. Goffi. 2021. Maximal L^q-regularity for parabolic Hamilton-Jacobi equations and applications to mean field games. *Annals of PDE* 7 (2): Paper No. 19, 40.
20. Cirant, M., and A. Porretta. 2021. Long time behavior and turnpike solutions in mildly non-monotone mean field games. *ESAIM Control Optimisation and Calculus of Variations* 27: Paper No. 86, 40.
21. Cirant, M., and D. Tonon. 2019. Time-dependent focusing mean-field games: the sub-critical case. *Journal of Dynamics and Differential Equations* 31 (1): 49–79.
22. Dieudonné, J. 1969. *Foundations of modern analysis*. New York-London: Academic Press. Enlarged and corrected printing. Pure and applied mathematics, vol. 10-I.
23. Dragoni, F., and E. Feleqi. 2018. Ergodic mean field games with Hörmander diffusions. *Calculus of Variations and Partial Differential Equations* 57 (5): Art. 116, 22.
24. Evans, L.C. 1990. *Weak convergence methods for nonlinear partial differential equations*, vol. 74. CBMS regional conference series in mathematics. Published for the conference board of the mathematical sciences. Washington, DC; by the American Mathematical Society, Providence, RI.
25. Feleqi, E., D. Gomes, and T. Tada. 2020. Hypoelliptic mean field games–a case study. *Minimax Theory and its Applications* 5 (2): 305–326.
26. Ferreira, R., and D. Gomes. 2018. Existence of weak solutions to stationary mean-field games through variational inequalities. *SIAM Journal on Mathematical Analysis* 50 (6): 5969–6006.
27. Ferreira, R., D. Gomes, and T. Tada. 2019. Existence of weak solutions to first-order stationary mean-field games with Dirichlet conditions. *Proceedings of the American Mathematical Society* 147 (11): 4713–4731.
28. Ferreira, R., D. Gomes, and T. Tada. 2021. Existence of weak solutions to time-dependent mean-field games. *Nonlinear Analysis* 212: Paper No. 112470, 31.
29. Gomes, D., H. Mitake, and K. Terai. 2020. The selection problem for some first-order stationary mean-field games. *Networks and Heterogeneous Media* 15 (4): 681–710.
30. Gomes, D., L. Nurbekyan, and E. Pimentel. 2015. *Economic models and mean-field games theory*. Publicações Matemáticas do IMPA. [IMPA Mathematical Publications]. Instituto Nacional de Matemática Pura e Aplicada (IMPA), Rio de Janeiro. 30° Colóquio Brasileiro de Matemática. [30th Brazilian Mathematics Colloquium].
31. Gomes, D., L. Nurbekyan, and M. Prazeres. 2017. One-dimensional stationary mean-field games with local coupling. *Dynamic Games and Applications*.
32. Gomes, D., S. Patrizi, and V. Voskanyan. 2014. On the existence of classical solutions for stationary extended mean field games. *Nonlinear Analysis* 99: 49–79.
33. Gomes, D., and E. Pimentel. 2015. Time dependent mean-field games with logarithmic nonlinearities. *SIAM Journal on Mathematical Analysis* 47 (5): 3798–3812.
34. Gomes, D., and E. Pimentel. 2016. Local regularity for mean-field games in the whole space. *Minimax Theory and its Applications* 01 (1): 065–082.
35. Gomes, D., E. Pimentel, and H. Sánchez-Morgado. 2015. Time-dependent mean-field games in the subquadratic case. *Communications in Partial Differential Equations* 40 (1): 40–76.
36. Gomes, D., E. Pimentel, and H. Sánchez-Morgado. 2016. Time-dependent mean-field games in the superquadratic case. *ESAIM Control Optimisation and Calculus of Variations* 22 (2): 562–580.

37. Gomes, D., E. Pimentel, and V. Voskanyan. 2016. *Regularity theory for mean-field game systems*. SpringerBriefs in mathematics. Cham: Springer.
38. Gomes, D., and H. Sánchez Morgado. 2014. A stochastic Evans-Aronsson problem. *Transactions of the American Mathematical Society* 366 (2): 903–929.
39. Gomes, D., and J. Saúde. 2014. Mean field games models–a brief survey. *Dynamic Games and Applications* 4 (2): 110–154.
40. Huang, M., P. E. Caines, and R. P. Malhamé. 2007. Large-population cost-coupled LQG problems with nonuniform agents: individual-mass behavior and decentralized ϵ-Nash equilibria. *IEEE Transactions on Automatic Control* 52 (9): 1560–1571.
41. Huang, M., R. P. Malhamé, and P. E. Caines. 2006. Large population stochastic dynamic games: closed-loop McKean-Vlasov systems and the Nash certainty equivalence principle. *Communications in Information and Systems* 6 (3): 221–251.
42. Lasry, J.-M., and P.-L. Lions. Jeux à champ moyen. II. Horizon fini et contrôle optimal. *Comptes Rendus Mathématique Académie des Sciences Paris* (10): 679–684.
43. Lasry, J.-M., and P.-L. Lions. 2006. Jeux à champ moyen. I. Le cas stationnaire. *Comptes Rendus Mathématique Académie des Sciences Paris* 343 (9): 619–625.
44. Lasry, J.-M., and P.-L. Lions. 2007. Mean field games. *Japanese Journal of Mathematics* 2 (1): 229–260.
45. Lions, P.L. 2007–2011. Collège de France course on mean-field games.
46. Mannucci, P., C. Marchi, and N. Tchou. 2021. First order Mean Field Games in the Heisenberg group: periodic and non periodic case. arXiv:2010.09279.
47. Mészáros, A. R., and F. J. Silva. 2018. On the variational formulation of some stationary second-order mean field games systems. *SIAM Journal on Mathematical Analysis* 50 (1): 1255–1277.
48. Pimentel, E., and V. Voskanyan. 2017. Regularity for second-order stationary mean-field games. *Indiana University Mathematics Journal* 66 (1): 1–22.
49. Porretta, A. 2014. On the planning problem for the mean field games system. *Dynamic Games and Applications* 4 (2): 231–256.
50. Porretta, A. 2015. Weak solutions to Fokker-Planck equations and mean field games. *Archive for Rational Mechanics and Analysis* 216 (1): 1–62.
51. Porretta, A. 2017. On the weak theory for mean field games systems. *Bollettino dell'Unione Matematica Italiana* 10 (3): 411–439.
52. Porretta, A., and M. Ricciardi. 2020. Mean field games under invariance conditions for the state space. *Communications in Partial Differential Equations* 45 (2): 146–190.
53. Voskanyan, V. K. 2013. Some estimates for stationary extended mean field games. *Dokl. Nats. Akad. Nauk Armen.* 113 (1): 30–36.

Simplifying Game Positions of the Green Hackenbush Combinatorial Game

Dantas Serra and João Luís Soares

Abstract One of the most striking results in the analysis of the Green Hackenbush game is the *Fusion Principle* [3, p. 189], a theorem that enables us to promptly determine the Sprague–Grundy value of any position by turning it into an *equivalent* position of the popular and well-studied Nim game. However, the constructive proof presented there is sketchy. This is the main reason behind this paper: to present a proof of the Fusion Principle without the omission of the main steps and details, building upon the work of [6].

1 Introduction

Green Hackenbush is a game played on a configuration of coloured green lines (though the chosen colouring is not important). These lines are connected to the ground, either directly by touching the ground or indirectly by being connected to another line that is connected to the ground. Players take turns cutting the line segments (e.g., erasing). When a line is cut, any remaining pieces that are no longer connected to the ground are also removed. The first player to be unable to move loses. The game can be tried online on [13], for example. The following image, taken from [9, p. 82], could be an initial configuration,

D. Serra · J. L. Soares (✉)
Departamento de Matemática, Universidade de Coimbra, Coimbra, Portugal
e-mail: jsoares@mat.uc.pt

D. Serra
e-mail: dantasp.serra@gmail.com

E. A. Pimentel and B. Toni (eds.), *Differential and Algorithmic Intelligent Game Theory*,
STEAM-H: Science, Technology, Engineering, Agriculture, Mathematics & Health,
https://doi.org/10.1007/978-3-031-97733-6_5

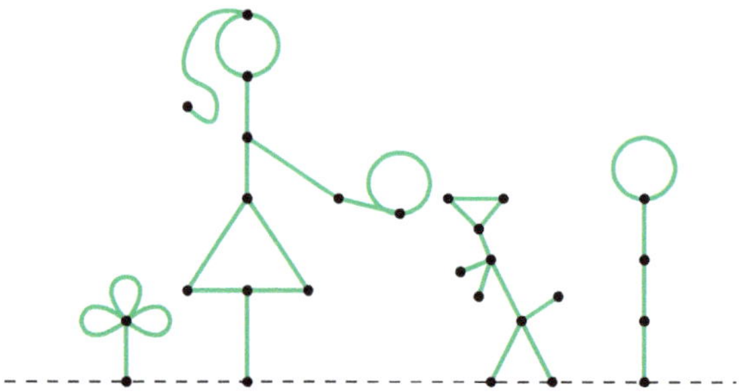

Hence, Green Hackenbush is an impartial (combinatorial) game under the normal termination rule. The game is analysed in [6], [5, p. 165] and [3, p. 189]. A partizan version of the game, known as *Red-and-Blue Hackenbush* is analysed in the latter reference. One of the most striking results in the analysis of the Green Hackenbush game is the *Fusion Principle* [3, p. 189], a theorem that enables us to promptly determine the Sprague–Grundy value of any position by turning it into an *equivalent* position of the popular and well-studied Nim game. However, the constructive proof presented there is sketchy. This is the main reason behind this paper: to present a proof of the Fusion Principle without the omission of the main steps and details, building upon the work of [6].

The paper is structured as follows. In Sect. 2, we recall the definition of combinatorial game, in both a broad and narrow sense, and make some historical remarks. In Sect. 3, we explain what is meant by solving a game, which is to classify the positions of the game into P and N. In Sect. 4, we introduce our main tool, the Sprague–Grundy Theorem and relate it to the Nim sum. In Sect. 5, we describe the Green Hackenbush game and present the easily solvable case that is similar to the Nim game. In Sect. 6, we prove the validity of the Colon Principle, which enables to reduce the no-cycles case to the Nim game. In Sect. 7, we introduce our main result that will be used in the proof of the validity of the Fusion Principle. Finally, in Sect. 8, we prove the validity of the Fusion Principle, which enables to reduce the general case, when cycles are present, to the Nim game.

2 What is a Combinatorial Game

In 1976, Conway [5] and, later, in 1982, Berlekamp, Conway and Guy [3], proposed a mathematical theory to analyse games without chance and without hidden information where two players take turns moving alternately. Some relevant related work was presented before, e.g., Bouton [4], Sprague [14], Grundy [8], but, it was not until 1982 that we could appreciate a complete and consistent theory, what we now call combinatorial game theory. Classical game theory [10]—related to economics and for which a nobel prize was won—expands combinatorial game theory in the sense that simultaneous moves and random moves (also called moves from nature) are also allowed. See [7, 9] for an elementary introduction, [11] for historical details. The topic is also relevantly surveyed in [1].

In a combinatorial game, two players take turns making moves alternately. A move takes a position of the game to another position. A position is a characterization of the game state in such a way that if the game is interrupted then it can be resumed the following day without loosing any information at all. No chance devices such as dice, spinners, or card deals are involved (i.e., there are no random moves), and each player is aware of all the past moves and positions at all times, i.e., it is a game of *perfect information*. No position can be repeated during the game and the game will end after a finite sequence of moves. In the end, at a so called *terminal position*, under normal play (assumed by omission) the winner is the player who made the last move or, under the *misére play* rule it is the other player who wins the game. The interested reader may find several combinatorial games in [2, 12]. Although a winner is often imposed on this combinatorial game definition, the possibility of ties generally poses no difficulty in the analysis. In any case, the description of a combinatorial game rules out the possibility of simultaneous moves and hidden moves.

If the rules of the game do not distinguish between each player's allowable moves at any given position, then we say that the (combinatorial) game is *impartial*. Otherwise, the game is *partizan*. In the Green Hackenbush game, there is no restriction on the line choice that each player will make. In the Red-and-Blue Hackenbush, where the lines are either colored Red or Blue, the player who initiates the game can only remove redish lines, while the other can only remove blueish lines. Hence, Red-and-Blue Hackenbush is partizan. Partizan games are harder to analyse.

For example, consider the following *subtraction game*, also refered as a *take-away game*. Two players alternately remove 1 or 4 chips from a pile that contains one million chips. This is an example of an impartial game under the normal rule. A different, though similar, game would be one in which the first player can remove 1 or 4 chips from the pile while the other can remove only 2 or 3. Now, this is a partizan game under the normal rule.

Of course, we can interpret a partizan game as an impartial game with an extra dimension in the characterization of a position (who is the next player to make a move). However, while this is certainly helpful in analysing games with few positions, it is not always clear whether this higher-dimensional interpretation of a game position allows for a solely analysis of the impartial version as a surrogate for any combinatorial game.

3 *P*-Positions and *N*-Positions

Returning to the two take-away games described before, we see that for the first game when the pile has seven chips or less, the positions 1, 3, 4 and 6 are winning for the **N**ext player (the player who sees this positions when it is his turn to move) while the positions 0, 2, 5 and 7 are winning for the **P**revious player (the player who just moved). Therefore, if the game had just started with seven chips, or if it had reached this stage of the game, the second player has the opportunity to win the game independently of how the first player moves.

In the second game, the partizan game, a similar analysis can be carried out. Now, for player A, the first to move, the positions $1, 2, 4$ and 5 are winning for the **N**ext player while the positions $0, 3, 6$ and 7 are winning for the **P**revious player. For player B, the second to play, the positions $2, 3, 5, 6$ and 7 are winning for the **N**ext player while the positions 0, 1 and 4 are winning for the **P**revious player. Therefore, if the game had just started with seven chips, B has the opportunity to win the game independently of how A moves.

In every combinatorial game the whole set of positions can be partitioned into P-positions and N-positions, where a P-position is a winning position for the **P**revious player and an N-position is a winning position for the **N**ext player. If the game is partizan then the partition may be distinct for both players. In fact, every combinatorial game can be understood as a game on a digraph $G \equiv (V, E)$ with no arc loops or dicycles—see [16] for a general reference on graphs and digraphs—where the elements of V, the vertices of G, are positions of the game and the elements of E, the arcs of G, are moves. At a given $x \in V$ (the position), the player selects an arc $xy \in E$ (the move) that will take the game to the position $y \in V$. For example, the impartial game described before is the game on the following digraph,

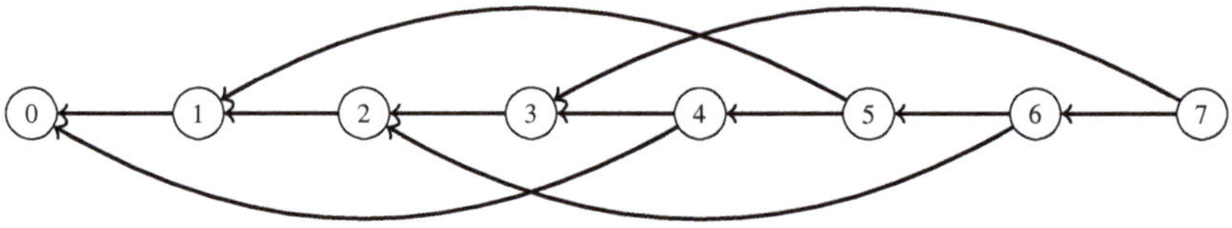

If the game is partizan then each arc is either colored red or blue. Player A, the first to move, can only select redish arcs while player B can only select blueish arcs. For example, the partizan take-away game is a game on the following digraph,

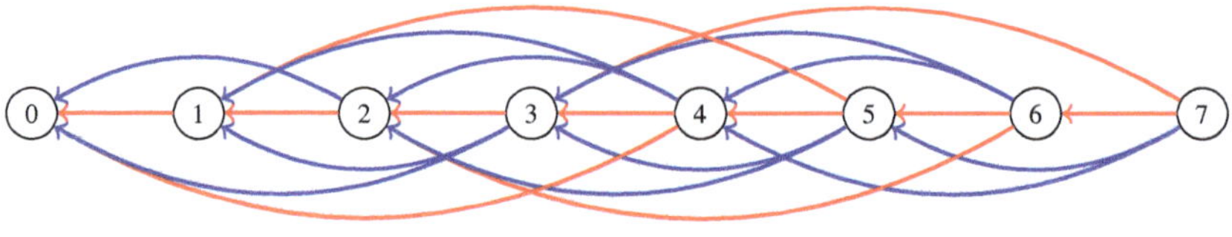

A dicycle is a sequence of consecutive arcs, all in the same direction, that describe a path from a given vertex to itself. In a game context, the existence of a dicycle implies the possibility of never ending. Therefore, by definition, the digraph that describes a combinatorial game has no dicycles.

A necessary and sufficient condition for a digraph G to have no dicycles is to have a topological sort of the vertices. A digraph G, having n vertices and m arcs, has a topological sort (of the vertices) if there exists a labeling function $l \colon V \to \{1, 2, \ldots, n\}$ such that $uv \in E$ implies $l(u) > l(v)$. In other words, if G can be drawn on the plane in such a way that all vertices are put on a line and all arcs are drawn from right to left. The reason for the equivalence is not very deep. If G has a topological sort, the existence of dicycle implies the existence of at least one arc uv in the dicycle in the *opposite* direction, a contradiction. If G has no dicycles then it must have a vertex v with outdegree equals to zero (e.g., the last vertex visited in a longest dipath). This is the vertex drawn furthest to the left on the line. Then, the same reasoning applies recursively on $G \setminus v$. In the end, a topological sort of the vertices of the initial G is evident. The running time of an algorithm that validates whether G has indeed no dicycles and finds a labeling function is $\mathcal{O}(n + m)$.

When a digraph $G \equiv (V, E)$ has a topological sort (of the vertices), any function g defined on V such that, for every $v \in V$, the value of $g(v)$ depends solely on the values of $g(w)$, for every w such that $l(w) < l(v)$, in a precise way, is well defined. Since, the fact that whether a given vertex v is and N- or P-position only depends on the whether vertices w with $l(w) < l(v)$ are N- or P-then every vertex of G can be classified as an N- or P-position without ambiguity.

Therefore, assuming that the labels of $V \equiv \{0, 1, \dots, n-1\}$ already reflect a topological sort on the digraph $G \equiv (V, E)$ then, in an impartial game under the normal termination rule, for every $i = 0, 1, \dots, n-1$, i is a P-position if there is no arc $ij \in E$ (in which case, i is terminal) or if, for every $ij \in E$, j is an N-position; i is an N-position if there is some $ij \in E$ such that j is a P-position (in which case ij is a *winning move*).

In a partizan game under the normal termination rule, for every $i = 0, 1, \dots, n-1$, i is a P-position for a given player if there is no arc $ij \in E$ (in which case, i is terminal) or if, for every $ij \in E$, j is an N-position for the other player; i is an N-position for a given player if there is some $ij \in E$ such that j is a P-position for the other player (in which case ij is a *winning move*). The application of these two algorithms to the games described above leads to the results exhbited in the following two tables

$$\begin{array}{cccccccc} 0 & 1 & 2 & 3 & 4 & 5 & 6 & 7 \\ \end{array}$$
$$\begin{bmatrix} P & N & P & N & N & P & N & P \end{bmatrix}$$

$$\begin{array}{c} \\ A \\ B \end{array}\begin{array}{c} \begin{array}{cccccccc} 0 & 1 & 2 & 3 & 4 & 5 & 6 & 7 \end{array} \\ \begin{bmatrix} P & N & N & P & N & N & P & P \\ P & P & N & N & P & N & N & N \end{bmatrix} \end{array}$$

The presence of ties can be handled in a similar way. Suppose that in a given impartial game under the normal termination rule there are some terminal positions in which none of the players is considered a winner. Then, the set of all positions can be classified into N-, P- and T-positions. The N-positions identify winning positions for the **N**ext player; the P-positions identify winning positions for the **P**revious player; and the T-positions identify positions which are neither winning for the **N**ext player or the **P**revious player. Algorithmically, they can be found in the following way. Assume that the labels of $V \equiv \{0, 1, \dots, n-1\}$ already reflect a topological sort on the digraph $G \equiv (V, E)$ and that $V' \equiv \{s, s+1, \dots, n-1\}$ are the nonterminal vertices. The vertices in $V \setminus V' \equiv \{0, 1, \dots, s-1\}$ are already classified as N, P and T, depending on the game rules. Then, for every $i = s, s+1, \dots, n-1$, i is P

if, for every $ij \in E$, j is N; i is T if, for every $ij \in E$, j is either N or T, and, for some $ij \in E$, j is T; i is N if, for some $ij \in E$, j is P. Of course, if, in particular, none of the terminal positions is T then none of whole set of positions is T either.

4 The Sprague–Grundy Theorem

Given a digraph $G \equiv (V, E)$ associated with a given impartial combinatorial game under a normal termination rule, the *Sprague–Grundy function of* G is a function $g\colon V \to \mathbb{N}_0$ defined by

$$g(x) = \text{mex}\{g(y) \colon xy \in E\},$$

where mex stands for "minimum excluded element from $\mathbb{N}_0$", i.e., if A is nonempty subset of $\mathbb{N}_0$ then $\text{mex}(A)$ is the smallest nonnegative integer that is not in A; if $A = \emptyset$ then $\text{mex}(A) = 0$. For example, for the impartial game above

$$g \equiv \begin{matrix} \scriptstyle 0 & \scriptstyle 1 & \scriptstyle 2 & \scriptstyle 3 & \scriptstyle 4 & \scriptstyle 5 & \scriptstyle 6 & \scriptstyle 7 \\ \end{matrix}$$
$$g \equiv \begin{bmatrix} 0 & 1 & 0 & 1 & 2 & 0 & 1 & 0 \end{bmatrix}$$

The Sprague–Grundy function is always well-defined because G has no dicycles. Moreover, by construction, for every $x \in V$, $g(x) = 0$ if and only if x is a P-position. In particular, whenever $g(x) > 0$, for every $xy \in E$, $g(y) = 0$ if and only if xy is a winning move.

The Sprague–Grundy function is a useful tool to compare positions in distinct games. Suppose that we have some position x in some game G and some position y in some other game H such that $g(x) = g(y) = p$. If $p = 0$ then both positions are P in their own game and, so, any move from those positions will take the game to a position with a positive Sprague–Grundy value. Now, if $p > 0$ then both positions are N and for each move in one game that will take the game to a position with a Sprague–Grundy value q smaller than p there is a corresponding move in the other game that will do the same. Later, we will refer to these two game positions as *equivalent*, but they are equivalent in this sense.

Every nonnegative integer a can be represented, and uniquely, as a sum of powers of two as follows,

$$a \equiv a_r \times 2^r + \cdots + a_1 \times 2^1 + a_0 \times 2^0 \equiv (a_r \ldots a_1 a_0)_2,$$

for some nonnegative r and some vector $\mathbf{a} \equiv (a_0, a_1, \ldots, a_r) \in \{0, 1\}^{r+1}$. For example,

$$11 = 8 + 2 + 1 = (1) \times 2^3 + (0) \times 2^2 + (1) \times 2^1 + (1) \times 2^0 \equiv (1011)_2.$$

Hence, we are able to define *Nim sum* of two nonnegative integers in the following way. If $a \equiv (a_r, \ldots, a_1, a_0)_2$ and $b \equiv (b_r, \ldots, b_1, b_0)_2$ then

$$a \oplus b = (a_r \oplus b_r, \ldots, a_1 \oplus b_1, a_0 \oplus b_0)_2 \equiv \sum_{i=0}^{r} (a_i \oplus b_i) 2^i, \tag{1}$$

where $a_i \oplus b_i = 1$ if $a_i \neq b_i$ and 0 otherwise, i.e., $0 \oplus 0 = 1 \oplus 1 = 0$ and $0 \oplus 1 = 1 \oplus 0 = 1$. For example, $3 \oplus 10 = (0011)_2 \oplus (1010)_2 = (1001)_2 \equiv 9$. Therefore, $3 \oplus 10$ is the sum of the powers of two that appear exactly once in both powers of two (or binary) representations. Note that $3 \oplus 10 = (2 + 1) \oplus (8 + 2) = 8 + 1$.

The operator "$\oplus$" is a group operator for $\mathbb{N}_0$ as one may show that $\oplus$ is associative, 0 is the identity element, and every element in $\mathbb{N}_0$ is its own inverse. The operator "$\oplus$" is also commutative. Moreover, the nim sum of two or more nonnegative integers is the sum of the powers of two that appear an odd number of times across the powers of two representations of all those numbers. For example,

$$26 \oplus 35 \oplus 12 \oplus 25 = (16 + 8 + 2) \oplus (32 + 2 + 1) \oplus (8 + 4) \oplus (16 + 8 + 1) = 32 + 8 + 4 = 44.$$

Lemma 1 below presents an alternative characterization of the Nim sum in terms of the mex function.

Lemma 1 *If* $a, b \in \mathbb{N}_0$ *then* $a \oplus b = \text{mex}\,(\{x \oplus b \colon x < a\} \cup \{a \oplus y \colon y < b\})$.

Proof If $x < a$ is such that $a \oplus b = x \oplus b$ then $a = x$, a contradiction. Similarly, if $y < b$ is such that $a \oplus b = a \oplus y$ then $b = y$, a contraction. Thus, $a \oplus b \notin A \equiv \{x \oplus b \colon x < a\} \cup \{a \oplus y \colon y < b\}$. Now, suppose $d \equiv \text{mex}(A) < a \oplus b$. If j is the position of the most significant digit in the binary expansion of $z \equiv d \oplus (a \oplus b)$ then the jth bit in the binary expansion of $a \oplus b$ is also "1". Thus: **(i)** the jth bit of the binary expansion of a is "1", or **(ii)** the jth bit of the binary expansion of b is "1", but not both. If **(i)** holds then $x \equiv a \oplus z < a$, and so $x \oplus b \equiv (a \oplus b) \oplus d \oplus (a \oplus b) \equiv d$, a contradiction; if **(ii)** holds then $y \equiv b \oplus z < b$, and so $a \oplus y \equiv (a \oplus b) \oplus d \oplus (a \oplus b) \equiv d$, another contradiction. Henceforth, we have concluded that $d = a \oplus b$. □

Consequently, if $z = a \oplus b$ then there is no $x < a$ such that $z = x \oplus b$, nor $y < b$ such that $z = a \oplus y$. For example, $3 \oplus 10 = \text{mex}\,(\{x \oplus 10 \colon x < 3\} \cup \{3 \oplus y \colon y < 10\}) = 9$ stems from

$\oplus$	0	1	2	3	4	5	6	7	8	9	10
0											10
1											11
2											8
3	3	2	1	0	7	6	5	4	11	10	**9**

Lemma 1 can be generalized to more than two sumands in an obvious way. The following two lemmas summarise a few useful properties of the nim sum.

Lemma 2 *If* $a, b \in \mathbb{N}_0$ *then:* **(i)** $a \oplus b$ *is odd if and only if* $a + b$ *is odd;* **(ii)** $(2a) \oplus (2b) = 2(a \oplus b)$*;* **(iii)** $a \oplus 1 = a + 1$ *if if* a *is even, or* $a - 1$ *if* a *is odd.*

Proof **(i)** Let $(a)_{10} \equiv (a_m a_{m-1} \cdots a_1 a_0)_2$, which is even if and only if $a_0 = 0$, and $(b)_{10} \equiv (b_m b_{m-1} \cdots b_1 b_0)_2$, which is even if and only if $b_0 = 0$. Then, $a + b$ is even if and only if the parity of a is equal to the parity of b, if and only if $a_0 = b_0$, if and only if the 0th digit of $(a \oplus b)_2$ is zero, if and only if $a \oplus b$ is even. **(ii)** If $(a)_{10} \equiv (a_m a_{m-1} \cdots a_1 a_0)_2$ and $(b)_{10} \equiv (b_m b_{m-1} \cdots b_1 b_0)_2$, then

$$(2a)_{10} \equiv (a_m a_{m-1} \cdots a_1 a_0 0)_2, \qquad (2b)_{10} \equiv (b_m b_{m-1} \cdots b_1 b_0 0)_2$$

so that the $(j+1)$th bit of $\big((2a) \oplus (2b)\big)_2$ matches the jth bit of $(a \oplus b)_2$, and the 0th bit of $\big((2a) \oplus (2b)\big)_2$ is zero. Therefore, $2(a \oplus b) = (2a) \oplus (2b)$. **(iii)** If a is a positive even integer, then $(a)_{10} \equiv (a_m a_{m-1} \cdots a_1 0)_2$, so that $(a \oplus 1)_{10} \equiv (a_m a_{m-1} \cdots a_1 1)_2 \equiv (a+1)_{10}$. If a is a positive odd integer, then $(a)_{10} \equiv (a_m a_{m-1} \cdots a_1 1)_2$, so that $(a \oplus 1)_{10} \equiv (a_m a_{m-1} \cdots a_1 0)_2 \equiv (a-1)_{10}$. □

Lemma 3 *If $a_1, a_2, \ldots, a_k$ are $k \geq 2$ non-negative integers, exactly p of which are odd, then*

$$a_1 \oplus \cdots \oplus a_k = \begin{cases} 2\left(\left\lfloor \frac{a_1}{2} \right\rfloor \oplus \cdots \oplus \left\lfloor \frac{a_k}{2} \right\rfloor\right) & \text{if } p \text{ is even,} \\ 2\left(\left\lfloor \frac{a_1}{2} \right\rfloor \oplus \cdots \oplus \left\lfloor \frac{a_k}{2} \right\rfloor\right) + 1 & \text{if } p \text{ is odd.} \end{cases}$$

Proof Without loss of generality, the odd numbers come before all the even ones, so that:

$$\begin{aligned} a_1 \oplus \cdots \oplus a_k &= \left((2\left\lfloor \tfrac{a_1}{2} \right\rfloor + 1) \oplus \cdots \oplus (2\left\lfloor \tfrac{a_p}{2} \right\rfloor + 1)\right) \oplus \left((2\left\lfloor \tfrac{a_{p+1}}{2} \right\rfloor) \oplus \cdots \oplus (2\left\lfloor \tfrac{a_k}{2} \right\rfloor)\right) \\ &= \left((2\left\lfloor \tfrac{a_1}{2} \right\rfloor \oplus 1) \oplus \cdots \oplus (2\left\lfloor \tfrac{a_p}{2} \right\rfloor \oplus 1)\right) \oplus \left((2\left\lfloor \tfrac{a_{p+1}}{2} \right\rfloor) \oplus \cdots \oplus (2\left\lfloor \tfrac{a_k}{2} \right\rfloor)\right) \\ &= \left((2\left\lfloor \tfrac{a_1}{2} \right\rfloor) \oplus \cdots \oplus (2\left\lfloor \tfrac{a_k}{2} \right\rfloor)\right) \oplus (1 \oplus \cdots \oplus 1) \\ &= 2\left(\left\lfloor \tfrac{a_1}{2} \right\rfloor \oplus \cdots \oplus \left\lfloor \tfrac{a_k}{2} \right\rfloor\right) \oplus \underbrace{(1 \oplus \cdots \oplus 1)}_{p \text{ times}}, \end{aligned}$$

where the second equality follows from **(iii)** in Lemma 2 and the fourth equality follows from **(ii)** in Lemma 2. The desired result follows from the fact that $1 \oplus \cdots \oplus 1$ is zero if p is even and one if p is odd. □

For example, $2 \oplus 3 \oplus 5 \oplus 2 \oplus 4 = 2(1 \oplus 1 \oplus 2 \oplus 1 \oplus 2)$ because there are 2 odd numbers. Moreover, $1 \oplus 1 \oplus 2 \oplus 1 \oplus 2 = 2(0 \oplus 0 \oplus 1 \oplus 0 \oplus 1) + 1$ because there are 3 odd numbers. Finally, since $0 \oplus 0 \oplus 1 \oplus 0 \oplus 1 = 0$ then $2 \oplus 3 \oplus 5 \oplus 2 \oplus 4 = 2$.

We will now explain how the nim sum is related to the Sprague–Grundy function. Given a number of impartial games $G_1, G_2, \ldots, G_k$, one can form a new impartial game $G \equiv G_1 + G_2 + \cdots + G_k$, played according to the following rules. A given position $\mathbf{x} \equiv (x_1, x_2, \ldots, x_k)$ is characterized by a given set of positions x_i in the games G_i. The two players alternate moves. A move consists in selecting any one of the games and making a legal move in that game, leaving all other games untouched.

The game proceeds until all of the games have reached a terminal position, when no more moves are possible. The player who made the last move is the winner. The game formed by combining games in this manner is called the *(disjunctive) sum* of the games $G_1, G_2, \ldots, G_k$.

One well-known example of a game that is defined in this way is the Nim game. There are k separate piles of chips. Two players take turns moving. Each move consists of selecting one of the piles and removing at least one chip from it. The winner is the player who removes the last chip. It turns out that the Sprague–Grundy value of each of component games is exactly the pile size. But, what is the Sprague–Grundy value of the overall pile set? The answer is addressed in the following theorem, from which Bouton's theorem on the Nim game [4] will stem as a corollary. Its proof is similar to the proof of Lemma 1.

Theorem 1 (Sprague Grundy Theorem [8, 14, 15]) *Let $G_1, G_2, \ldots, G_k$ denote a set impartial games. The Sprague–Grundy function of the (disjunctive) sum game $G \equiv G_1 + G_2 + \cdots + G_k$ at a position $\mathbf{x} \equiv (x_1, x_2, \ldots, x_k)$ is*

$$g(\mathbf{x}) = g_1(x_1) \oplus g_2(x_2) \oplus \cdots \oplus g_k(x_k),$$

where g_i denotes the Sprague–Grundy function of the game G_i.

Therefore, if we know the Sprague–Grundy function of each of the games G_i at any given position x_i, we may immediately realize whether $\mathbf{x} \equiv (x_1, x_2, \ldots, x_k)$ is P or N. In fact, if $g(\mathbf{x}) = 0$ then $\mathbf{x}$ is P. If, on the other hand, $g(\mathbf{x}) > 0$ then, from the definition of the Sprague–Grundy function, there is some move that takes $\mathbf{x}$ to some P-position $\mathbf{y}$, i.e., one in which $g(\mathbf{y}) = 0$. How do find $\mathbf{y}$? From Lemma 1, or its generalization, if $k > 2$, for one of the games G_j there is a move that takes it from position x_j to some position y_j so that the position $\mathbf{y}$ in G has Sprague–Grundy value equal to zero.

For example, suppose that the set of pile sizes in the Nim game is $\mathbf{x} \equiv (26, 35, 12, 25)$. Since $g_i(x_i) \equiv x_i$ then $g(\mathbf{x}) = 26 \oplus 35 \oplus 12 \oplus 25 \equiv 32 + 8 + 4 = 44$. Hence, $\mathbf{x}$ is N. What is the optimal move? Simply identify some game G_j for which $g_j(x_j)$ shares the presence of the largest power of two in 44. Then, make the optimal move in G_j. In this case, there is a unique possibility, to select the pile that has $35 \equiv 32 + 2 + 1$ chips and remove a number of chips that makes the new pile size be $(2 + 1) + (8 + 4) \equiv 15$, i.e., remove 20 chips. The new position $\mathbf{y} \equiv (26, 15, 12, 25)$ is P. This is basically why we loosely say that *every impartial game position is equivalent to a nim game position.*

5 The Green–Hackenbush Game

Any position of the Green Hackenbush game is characterized by a finite number of independent graphs, each of which has one or more vertices that *attaches* the graph to the ground. We call these rooted graphs. The ground is denoted in the figures that follow by a dotted line. For simplificaton purposes, we will assume that no more than one vertex serves as *root* or *ground*. This can be assumed without loss of generality since *sticking* together the vertices attached to the ground does not change the game, as the following two examples show,

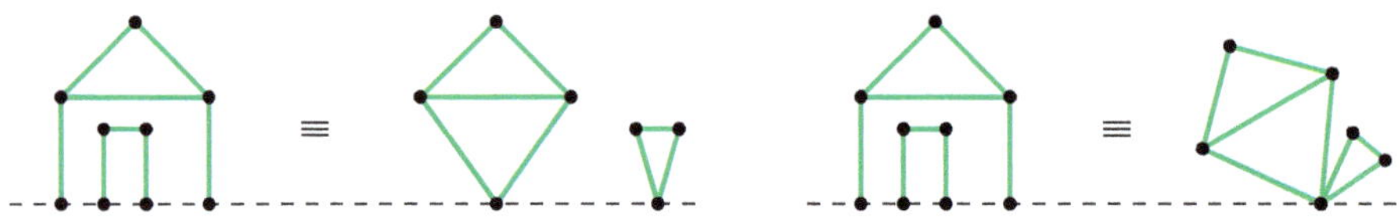

So, a rooted graph is a pair (G, r), where G is an undirected graph and a vertex $r \in V$ such that every other vertex of G is connected to r along some path in G. A move consists in choosing one rooted graph (G_i, r_i) and removing one edge from it. After each play, every vertex that becomes disconnected from the ground is removed from the game. All edges incident to a removed vertex are also removed. The winner is the player who makes the last move.

Hence, the Green Hackenbush game can be understood as a (disjunctive) sum of several Green Hackenbush games, each of which consisting of a single rooted graph. Thus, to compute the Sprague–Grundy value at any position we may use the Sprague–Grundy Theorem.

Consider the simplest type of rooted graph that can appear when playing Green Hackenbush, a bamboo stalk. A bamboo stalk with n segments is a linear graph of n edges with the bottom of these n edges rooted to the ground. A single bamboo stalk of n segments can be moved into a bamboo stalk of any smaller number of segments from $n - 1$ to 0. So, a single bamboo stalk of n segments is equivalent to a nim pile of n chips. Playing a (disjunctive) sum of of bamboo stalks is equivalent to playing Nim. From Theorem 1, we see that the Sprague–Grundy value of the position $\mathbf{x}$ below is $(2 + 1) \oplus (2) \oplus (2 + 1) \oplus (1) = 2 + 1 \equiv 3$, Thus, $\mathbf{x}$ is an N-position. One of the optimal moves (there are three of them) is to remove the edge in the bottom of the first bamboo stalk.

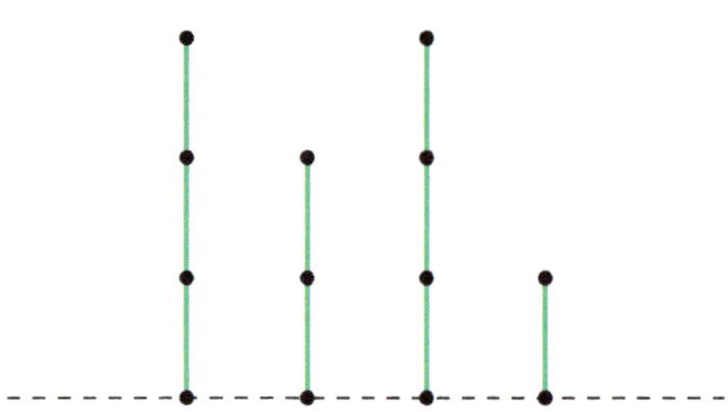

6 The Colon Principle

From now on, our primary goal will be to simplify any position in the Green–Hackenbush game into a position that consists of several bamboo stalks. The simplification should keep the Sprague–Grundy value. The *Colon Principle* is the technique that allows for the simplification of a rooted tree into a bamboo stalk. A rooted tree is a rooted graph with no cycles. I.e., from every vertex in the graph there is a unique path to the root. It works as follows recursively: *when branches of a rooted tree come together at a vertex, one may replace the branches by a non-branching stalk of length equal to their Nim sum.* We will show the validity of this principle in Proposition 1 below.

Proposition 1 (validity of the Colon Principle) *Let $G \equiv (V, E)$ be a graph rooted at r and let H_1, H_2 be two distinct graphs rooted at $v \in V$, their unique intersection with G. If their Sprague Grundy value is the same then the Sprague Grundy of the graphs $G \cup H_1$ and $G \cup H_2$, both rooted at r, is also the same.*

Proof Consider the configuration of the Green–Hackenbush game made of two distinct rooted graphs, the rooted graphs $(G_1 \equiv G \cup H_1, r)$ and a copy of $(G \cup H_2, r)$, denoted $(G_2 \equiv G \cup H_2, r')$, as the following figure suggests,

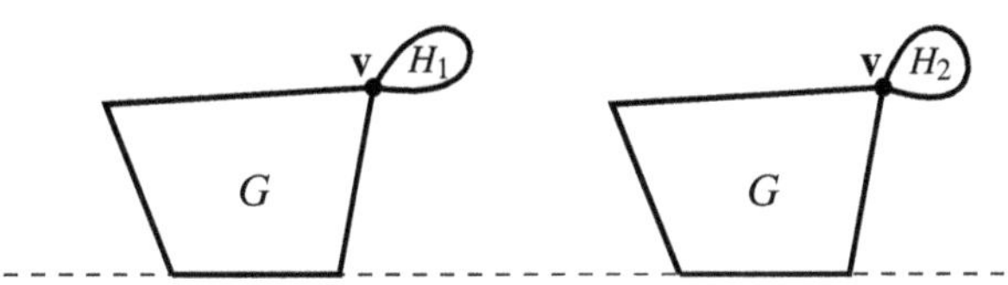

We will show that its Sprague–Grundy value is 0 by showing that it is a P-position. Player I starts by removing an edge from G_1, without loss of generality. If the move corresponds to removing an edge of G then Player II, in his turn, can remove the same edge in G_2. After these two moves, the two new rooted graphs have the same properties they had before. Note that, in particular, both H_1, H_2 may have disappeared.

If Player I's move corresponds to removing an edge of H_1 then, after the removal, the rooted graph (H_1, v) becomes (H_1', v). If the Sprague–Grundy value of (H_1', v) is lower than that of (H_2, v) then, from the definition of the Sprague–Grundy function, there is move on the rooted graph (H_2, v) that takes it to some other graph (H_2', v) in such a way that both rooted graphs (H_1', v) and (H_2', v) have the same Sprague–Grundy value. Otherwise, from the definition of the Sprague–Grundy function, the Sprague–Grundy value of (H_1', v) is higher than that (H_2, v). Then, again from the definition of the Sprague–Grundy function, there is a move on the rooted graph (H_1', v) that takes it to some other graph (H_1'', v) in such a way that both rooted graphs (H_1'', v) and (H_2, v) have the same Sprague–Grundy value.

In any case, we end up with two new rooted graphs that have the same properties they had two moves before. By playing this way, Player II always has a response to every move made by Player I, proving that the initial position is a P-position. Hence, from Sprague–Grundy's Theorem, the Sprague Grundy value of the graphs $G \cup H_1$ and $G \cup H_2$, both rooted at r, is the same. □

In particular, the Colon Principle can be applied recursively to any rooted tree. Any tree ramification, starting with the ones further way from the root, can be replaced by a bamboo stalk containing as many edges as the Sprague–Grundy value of the ramification. After a finite number of such steps we end up with a single bamboo stalk. In the following game, we start by replacing the branch made of three bamboo stalks by a single bamboo stalk of size $1 \equiv 1 \oplus 1 \oplus 1$. In the second step, we replace the ramification made of two bamboo stalks by a single bamboo stalk of size $3 \equiv 2 \oplus 1$.

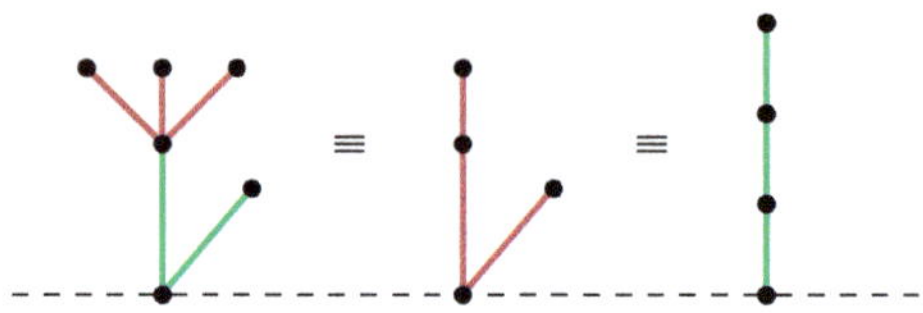

Corollary 1 *(Parity Principle) At a rooted tree, the parity of the Sprague–Grundy value and the number of edges is the same.*

Proof Any tree can be fully simplified to a bamboo stalk by applying the Colon Principle to each ramification that consists only of bamboo stalks. From **(i)** in Lemma 2, the Nim sum preserves parity. Therefore, each simplification does not change the parity of the number of edges. The final simplification yields a bamboo stalk whose length is equal to the Sprague–Grundy value. □

An immediate consequence of the Parity Principle (Corollary 1) is that any rooted tree with an odd number of edges is an N-position. Moreover, it is an easy exercise to conclude from the Parity Principle that, for any game $G = H_1 + \cdots + H_n$, where each H_i is a rooted tree, the Sprague–Grundy value of G has the same parity as the number of edges in it, since

$$g(G) = g(H_1) \oplus \cdots \oplus g(H_n)$$

and, by Lemma 2, the nim sum preserves parity.

A special case arises when the rooted tree is a *simple rooted tree*. A simple rooted tree is a rooted tree (T, r) whose vertices have degree 3 or less, and such that all vertices of T with degree 3 lie on a unique path starting at r—called the *main path*. The branches (bamboo stalks) leaving the main path are called *secondary trees*. The rooted graph below on the left is not a simple tree because it has vertices of degree 4. The rooted graph on the right is a simple rooted tree.

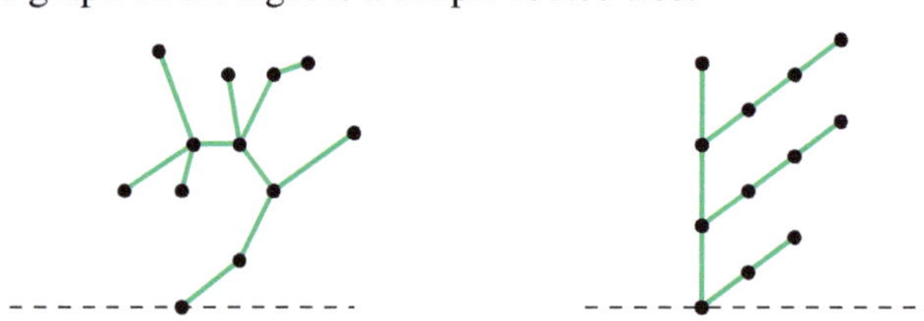

Typically, simple rooted trees are represented with the main path drawn vertically and the secondary trees drawn diagonally. A bamboo stalk is a simple rooted tree with no secondary trees. More examples are as follows,

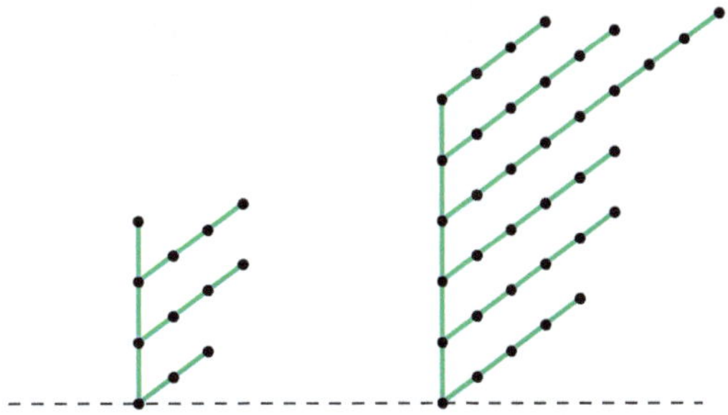

Thus, a simple rooted tree can be identified by the lengths of its secondary trees from top to bottom. The previous examples can be represented by the vectors $(0, 3, 3, 2)$ and $(3, 5, 8, 5, 5, 4)$, respectively. In general, a simple rooted tree position whose main path has $k + 1$ vertices is identified by some vector $(a_0, a_1, \ldots, a_k)$, where a_0 is the length of the secondary tree furthest away from the root and a_k is the length of the secondary tree connected to the root.

From the Colon Principle, we can compute the Sprague–Grundy value of a simple rooted tree by evaluating a certain algebraic expression.

Lemma 4 (simple rooted tree SG value) *The Sprague–Grundy value of (T, r), a simple rooted tree identified by some vector $\mathbf{a} \equiv (a_0, a_1, \ldots, a_k)$, is*

$$f(\mathbf{a}) = a_0 + 1 \oplus a_1 + \cdots \oplus a_{k-1} + 1 \oplus a_k,$$

where the operators "+" and "⊕" have the same precedence.

Proof The top secondary tree, together with the last edge from the main path, has length $a_0 + 1$. Applying the Colon Principle to simplify the trees corresponds to replacing them by the bamboo stalk with the Nim sum $a_0 + 1 \oplus a_1$. After applying the Colon simplification, the top secondary tree has length $a_0 + 1 \oplus a_1$. By repeating this process until the tree is converted into a bamboo stalk, the full algebraic expression is obtained. □

Algorithmically, if a simple rooted tree (T, r) is identified by some vector $\mathbf{a} \equiv (a_0, a_1, \ldots, a_k)$ then $f(\mathbf{a}) = f_k$, where f_k is obtained though the computation of $f_i = f_{i-1} + 1 \oplus a_i$, for $i = 1, 2, \ldots, k$, starting from $f_0 = a_0$. For example, the fact that $f(4, 6, 3, 8, 0) = 1$ and $f(4, 5, 8, 5, 5, 4) = 18$ follows from the intermediate calculations presented in the following tables,

i	a_i	f_i
$i = 0$	4	4
1	6	3
2	3	7
3	8	0
4	0	1

i	a_i	f_i
$i = 0$	4	4
1	5	0
2	8	9
3	5	15
4	5	21
5	4	18

7 A Reduced Rooted Tree

In this section, we will explain how to build a simple rooted tree with roughly half the edges of another simple rooted tree in such a way that the Sprague–Grundy values of both are related. The new simple tree will be constructed in four steps. Let (T, r) be a simple rooted tree identified by some vector $(a_0, a_1, \ldots, a_k)$. Then:

Step 1: Label the main path edges with the letters E and O, from the top down, according to the following rules: **(i)** label the furthest edge away from the root with E if a_0 is even or O if a_0 is odd; **(ii)** Two adjacent edges will have the same label if and only if there is an odd-length bamboo stalk attached to the common vertex.

Step 2: Halve the length of every secondary tree from a_i to $\lfloor a_i/2 \rfloor$.

Step 3: Shrink every E-marked edge from the main path into a single vertex.

Step 4: After Step 3, the main path may have vertices with more than one bamboo stalk attached. Therefore, use the Colon Principle to simplify all secondary trees attached to the same vertex of the main path into a single bamboo stalk, turning the tree again into a simple tree.

We will call the simple rooted tree obtained after these four steps a *reduced tree* and will denote it by rT, omitting the root node for clarity. For example, consider the simple rooted tree below, identified by $(1, 2, 3, 2, 1)$. Step 1: since $a_0 = 1$ is odd, the labeling starts with O; since $a_1 = 2$ is even, the next label is different, i.e., E; since $a_2 = 3$ is odd, the next label is the same, i.e., E; since $a_3 = 2$ is even, the next label is different, i.e., O. Step 2: by halving the secondary trees we have that $\lfloor 1/2 \rfloor = 0$, $\lfloor 2/2 \rfloor = 1$, $\lfloor 3/2 \rfloor = 1$, $\lfloor 2/2 \rfloor = 1$ and $\lfloor 1/2 \rfloor = 0$ are the sizes of the new secondary trees. Step 3: the two edges marked with E are shrinked into one single vertex. Step 4: after the shrinkage, three branches emerge from the second node, top down. The Colon Principle allows for the final simplification for $1 \oplus 1 \oplus 1 = 1$. The algebraic description above is represented in the figure below,

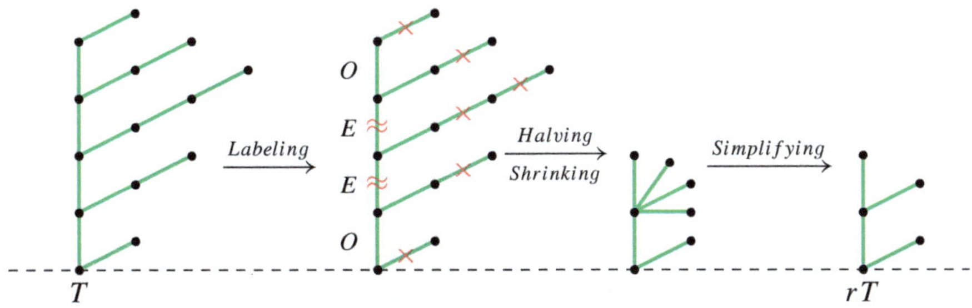

Note that since the reduced tree rT that stems from T is still a simple tree, the reduction process can be applied again, and again. We will denote by $r^m T$ the result of applying the reduction procedure recursively m times starting from a simple rooted tree T.

Now, in order to relate the Sprague–Grundy value of T and the Sprague–Grundy value of rT, let's interpret the labeling rule in an alternative way. Let T be a simple rooted tree identified by $\mathbf{a} \equiv (a_0, a_1, \ldots, a_k)$. Let $\mathbf{n} \equiv (n_0, n_1, \ldots, n_k)$ be a vector whose generic component n_i equals the number of even components in $\mathbf{a}_i \equiv (a_0, a_1, \ldots, a_i)$. For example, if $\mathbf{a} \equiv (1, 2, 3, 2, 1)$ then $\mathbf{n} \equiv (0, 1, 1, 2, 2)$. The main path of a simple tree with $k+1$ vertices has k edges—call them $e_1, e_2, \ldots, e_k$, from top to bottom. The labeling of the first step of the reduction process is equivalent to labeling edge e_i with E if n_{i-1} is odd and with O if n_{i-1} is even. In the previous example the labeling would produce $OEEO$, as the figure confirms. The next theorem relates the Sprague–Grundy value of both simple rooted trees, T and rT.

Theorem 2 *Let T be a simple rooted tree identified by $\mathbf{a} \equiv (a_0, a_1, \ldots, a_k)$. The Sprague–Grundy value of T is*

$$f(\mathbf{a}) = \begin{cases} 2g_k & \text{if } n_k \text{ is odd,} \\ 2g_k \oplus 1 & \text{if } n_k \text{ is even,} \end{cases} \tag{2}$$

where g_k, the Sprague–Grundy value of rT, is obtained through

$$g_i = \begin{cases} g_{i-1} \oplus \lfloor a_i/2 \rfloor, & \text{if } n_{i-1} \text{ is odd,} \\ g_{i-1} + 1 \oplus \lfloor a_i/2 \rfloor, & \text{if } n_{i-1} \text{ is even.} \end{cases}$$

for every $i = 1, 2, \ldots, k$, starting from $g_0 = \lfloor a_0/2 \rfloor$.

Proof By mathematical induction on k. If $k = 0$, $\mathbf{a} \equiv (a_0)$. If a_0 is even, then n_0 is odd and $f(\mathbf{a}) \equiv a_0 = 2\lfloor a_0/2 \rfloor = 2g_0$; if a_0 is odd, then n_0 is even and $f(\mathbf{a}) \equiv a_0 = 2\lfloor a_0/2 \rfloor + 1 = 2g_0 + 1$. Therefore, (2) is true when $k = 0$.

Suppose, by the induction hypothesis, that (2) is true for a given $k \geq 0$, and let $\mathbf{a} \equiv (a_0, a_1, \ldots, a_k, a_{k+1})$ be a vector with $(k+1)+1$ non-negative integer components. **(i)** If a_{k+1} is even and n_k is odd, then, by the induction hypothesis and Lemma 3,

$$f(\mathbf{a}) = (2g_k) + 1 \oplus a_{k+1} \equiv (2g_k + 1) \oplus 2\lfloor a_{k+1}/2 \rfloor$$
$$= 2(g_k + \lfloor a_{k+1}/2 \rfloor) + 1 \equiv 2g_{k+1} + 1.$$

Thus, in this case, (2) follows from the fact that n_{k+1} is even. **(ii)** If a_{k+1} is even and n_k is even, then, by the induction hypothesis and Lemma 3,

$$f(\mathbf{a}) = (2g_k + 1) + 1 \oplus a_{k+1} \equiv 2(g_k + 1) \oplus 2\lfloor a_{k+1}/2 \rfloor$$
$$= 2(g_k + 1 + \lfloor a_{k+1}/2 \rfloor) \equiv 2g_{k+1}.$$

Thus, in this case, (2) follows from the fact that n_{k+1} is odd. **(iii)** If a_{k+1} is odd and n_k is odd, then, by the induction hypothesis and Lemma 3,

$$f(\mathbf{a}) = (2g_k) + 1 \oplus a_{k+1} \equiv (2g_k + 1) \oplus 2\lfloor a_{k+1}/2 \rfloor + 1$$
$$= 2(g_k + \lfloor a_{k+1}/2 \rfloor) \equiv 2g_{k+1}.$$

Thus, in this case, (2) follows from the fact that n_{k+1} is odd. Finally, **(iv)** If a_{k+1} is odd and n_k is even, then, by the induction hypothesis and Lemma 3,

$$f(\mathbf{a}) = (2g_k + 1) + 1 \oplus a_{k+1} \equiv 2(g_k + 1) \oplus 2\lfloor a_{k+1}/2 \rfloor + 1$$
$$= 2(g_k + 1 + \lfloor a_{k+1}/2 \rfloor) + 1 \equiv 2g_{k+1} + 1.$$

Thus, in this case, (2) follows from the fact that n_{k+1} is even, and the first part of the proof is complete.

It remains to be shown that the Sprague–Grundy value of rT is g_k. Let's proceed by induction in k, the length of T's main path.

If $k = 0$, $T = (a_0)$, and the reduction process halves the only secondary tree in it, then rT consists of one bamboo stalk of length $\lfloor a_0/2 \rfloor$, and $g(rT) = \lfloor a_0/2 \rfloor = g_0$, ending the case $k = 0$.

Suppose, by the induction hypothesis, that for any simple tree T' with k edges on the main path, $g(rT') = g_k$. Let $T = (a_0, a_1, \ldots, a_k, a_{k+1})$ be a simple rooted tree whose main path has $k + 1$ edges.

The main argument for the induction step is that the simple rooted tree $T' = (a_0, a_1, \ldots, a_k)$ corresponds to the simple tree attached to the rooted edge e_{k+1} of the main path of T. Also, during the reduction step, the main path is labeled according to the vector $n = (n_0, n_1, \ldots, n_k, n_{k+1})$. Since each component n_i is defined by $(a_1, \ldots, a_i)$, the labeling of T' follows from the vector $n' = (n_0, n_1, \ldots, n_k)$, meaning that the reduction process on T' is the same as if it were attached to the ground (allowing us to use the induction hypothesis because T' has a k-length main path).

(i) If n_k is odd, the edge e_{k+1} must be shrunk during the reduction process to obtain rT, and the reduced tree rT' will be attached to the ground. The secondary tree given by a_{k+1} will be halved into $\lfloor a_{k+1}/2 \rfloor$.

By the Colon Principle 1 and the induction hypothesis, rT' can be replaced by a bamboo stalk of length $g(rT') = g_k$. Finally, the Sprague–Grundy value is obtained by taking the nim-sum of the g_k-length bamboo stalk and the $\lfloor a_{k+1}/2 \rfloor$-length bamboo stalk, that is,

$$g(rT) = g_k \oplus \lfloor a_{k+1}/2 \rfloor = g_{k+1}.$$

(ii) On the other hand, if n_k is even, the edge e_{k+1} is not shrunk. Again, the Colon Principle and the induction hypothesis allow us to replace rT' with a g_k-length bamboo stalk. However, since e_{k+1} was not shrunk, the total length of the bamboo stalk will be $g_k + 1$. Once again, the Sprague–Grundy value of rT must be obtained by taking the nim-sum of $g_k + 1$ with the halved secondary tree $\lfloor a_{k+1}/2 \rfloor$, that is,

$$g(rT) = g_k + 1 \oplus \lfloor a_{k+1}/2 \rfloor = g_{k+1}.$$

This concludes the proof. □

As an illustration of Theorem 2, consider a simple rooted tree T identified by $\mathbf{a} = (4, 5, 8, 5, 5, 4)$. Hence, $k = 5$ and, as have seen before, $f(\mathbf{a}) = 18$. Both T and rT are represented below,

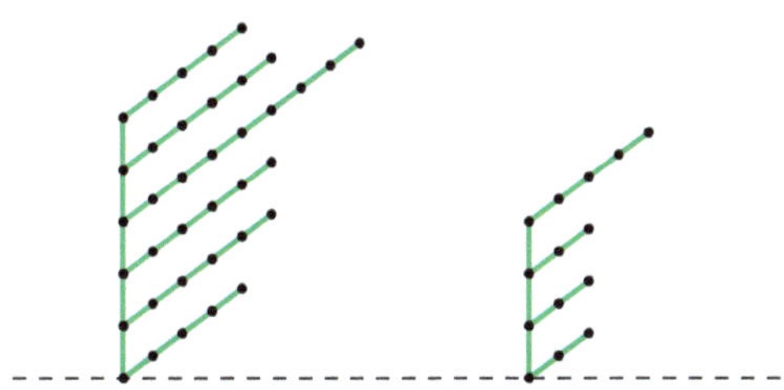

Since $\mathbf{n} = (1, 1, 2, 2, 2, 3)$, the calculation of $g_5 = 9 \equiv 18/2$, because $n_5 \equiv 3$ is odd, stems from the following calculations

$$g_5 \equiv \overbrace{\overbrace{\overbrace{\underbrace{\underbrace{\underbrace{2}_{g_0} \oplus 2}_{g_1} \oplus 4}_{g_2} + 1 \oplus 2}^{g_3} + 1 \oplus 2}^{g_4} + 1 \oplus 2}^{g_5}$$

whose intermediate values are presented in the following table,

	$\lfloor a_i/2 \rfloor$	n_i	$g_i - g_{i-1}$	g_i
$i=0$	2	1	2	2
1	2	1	$\oplus 2$	0
2	4	2	$\oplus 4$	4
3	2	2	$+1 \oplus 2$	7
4	2	2	$+1 \oplus 2$	10
5	2	3	$+1 \oplus 2$	9

Note that the position of the "+1" terms in the algebraic expression of g_5 corresponds to the edges that were not shrunk, the edges that received the label O.

Corollary 2 **(i)** *If T is a simple rooted tree, then*

$$g_{SG}(T) = \begin{cases} 2g_{SG}(rT), & \text{if } T \text{ has even number of edges,} \\ 2g_{SG}(rT) \oplus 1, & \text{if } T \text{ has odd number of edges.} \end{cases}$$

(ii) *If $T_1, \ldots, T_\ell$ are distinct simple rooted trees whose total number of edges is even then*

$$g_{SG}(T_1 + \cdots + T_\ell) = 2g_{SG}(rT_1 + \cdots + rT_\ell)$$

Proof From Theorem 2, $g_{SG}(T) = 2g_{SG}(rT)$ or $g(T) = 2g(rT) + 1$. By the Parity Principle (Corollary 1), T has an even number of edges if and only if $g_{SG}(T)$ is even. Thus, **(i)** is proved. Now, we prove **(ii)**. Since the total number of edges in $T_1, \ldots, T_\ell$ is even, by the Parity Principle, $g_{SG}(T_1 + \cdots + T_\ell)$ is also even. From the Sprague Grundy theorem,

$$g_{SG}(T_1 + \cdots + T_\ell) = g_{SG}(T_1) \oplus \cdots \oplus g_{SG}(T_\ell).$$

Therefore, this sum has even number odd sumands. From **(i)**, and the commutativity of the Nim sum,

$$g_{SG}(T_1) \oplus \cdots \oplus g_{SG}(T_\ell) = 2g_{SG}(rT_1) \oplus \cdots \oplus 2g_{SG}(rT_\ell) \oplus \underbrace{1 \oplus \cdots \oplus 1}_{\text{even amount}} = 2g_{SG}(rT_1) \oplus \cdots \oplus 2g_{SG}(rT_\ell)$$

Now, from **(ii)** in Lemma 2 and from the Sprague Grundy Theorem,

$$2g_{SG}(rT_1) \oplus \cdots \oplus 2g_{SG}(rT_\ell) = 2\left(g_{SG}(rT_1) \oplus \cdots \oplus g_{SG}(rT_\ell)\right) = 2g_{SG}(rT_1 + \cdots + rT_\ell).$$

□

An immediate consequence of **(i)** in Corollary 2 is that for a simple rooted tree T identified by $\mathbf{a} \equiv (a_0, \ldots, a_k)$, $g_{SG}(T)$ is even if and only if n_k is odd, meaning that $g_{SG}(T)$ and n_k have different parities. On the other hand, **(ii)** in Corollary 2 gives us an alternative way to compute the Sprague Grundy value of any sum of distinct simple rooted trees whenever overall they have an even amount of edges. Moreover, since the reduction process can be applied any number of times, the following corollary naturally holds.

Corollary 3 (Even reduction) *Let $T_1, \ldots, T_\ell$ denote a collection of distinct simple rooted trees. If $r^i T_1 + \cdots + r^i T_\ell$ has an even number of edges, for every $i = 0, 1, \ldots, m-1$, then*

$$g(T_1 + \cdots + T_\ell) = 2^m g(r^m T_1 + \cdots + r^m T_\ell)$$

Proof Immediate from applying Corollary 2 m times. □

8 The Fusion Principle

The Fusion Principle allows us to deal with cycles in Green Hackenbush positions. The proof of its validity is more difficult than that of the Colon Principle. The Fusion Principle is as follows: *two (distinct) vertices belonging to some cycle may be fused, i.e., merged into a single vertex and, eventually, any edge linking them together is converted into a loop at the new vertex*. The fusion process does not change the number of edges but it decreases by one the number of vertices. The following figure exemplifies the result from fusing two distinct pairs of vertices in a cycle. The pairs are coloured grey.

The main idea of the proof of the validity of the Fusion Principle is to transform geometric properties into algebraic ones in order to simplify the calculation of the Sprague–Grundy value of certain special Green Hackenbush positions. Our goal is to show that fusing two vertices does not change the Sprague–Grundy value. This is equivalent to proving that no graph exists in which this operation alters the game's value. For organizational purposes, we define the following concept.

Definition 1 (*Minimal Non-Fusable Graph*) A rooted graph (or game position) G containing a cycle is *non-fusable* if it is impossible to fuse any two vertices without changing the Sprague–Grundy value. A nonfusable G is *minimal non-fusable* (MNF for short) if it is minimal in the number of edges and, among these, minimal in the number of vertices.

Indeed, our goal will be to show that no MNF Green Hackenbush position exists. To do so, let's examine the properties that such a graph must satisfy.

Lemma 5 (limited paths) *If G is MNF, no two (distinct) vertices v, w are connected by, at least, three or more edge-disjoint paths.*

Proof By contradiction, suppose that there are two distinct vertices v, w connected by at least three edge-disjoint paths. Since these vertices are in a cycle, we can apply fusion. Let H denote the graph obtained after fusing v, w. Since G is MNF, $g_{\text{SG}}(G) \neq g_{\text{SG}}(H)$. Therefore, from the Sprague Grundy Theorem, $g_{\text{SG}}(G + H) = g_{\text{SG}}(G) \oplus g_{\text{SG}}(H) > 0$, i.e., $G + H$ is a an N-position.

Now, Player I makes his optimal move, which is to to take some edge from either G or H. Then, Player II should remove the very same edge, but in the opposite graph. After two moves, the position $G + H$ is changed to $G' + H'$ and $g_{\text{SG}}(G' + H') > 0$. Both graphs, G' and H' have the same number of edges, at most one less the number of edges of both G, H.

Since v, w are connected in G by, at least, three or more edge-disjoint paths then v, w are still in a cycle in G'. Let G'' denote the graph obtained from G' after fusing

v, w. Both graphs G'' and H' are the same (this can be easily shown considering a few cases separately) and, since G is MNF, $g_{\text{SG}}(G') = g_{\text{SG}}(G'')$. Thus, $g_{\text{SG}}(G' + H') = g_{\text{SG}}(G') \oplus g_{\text{SG}}(H') = g_{\text{SG}}(G'') \oplus g_{\text{SG}}(H') = 0$, a contradiction. □

Lemma 6 (grounded cycle) *If G is an MNF, every cycle in G must contain the ground vertex.*

Proof Let C be an arbitrary cycle in G. By contradiction, suppose that the ground vertex r does not belong to C. Since, from Lemma 5, there are no two (distinct) vertices v, w connected by, at least, three or more edge-disjoint paths, then there exists some vertex x in C that is a vertex-cut of G.

Now, on one side, let H denote the maximal subgraph of G that contains x and all the vertices that are on the same side as C and consider H rooted at x. On the other side, let G' denote the maximal subgraph of G that contains x and all the vertices that are on the same side as r and consider G' rooted at r. Since G is MNF, we can fuse two of the vertices of C in H and obtain some rooted graph H' such that $g_{\text{SG}}(H) = g_{\text{SG}}(H')$. From the Colon Principle,

$$g_{\text{SG}}(G) \equiv g_{\text{SG}}(G' \cup H) = g_{\text{SG}}(G' \cup H'),$$

i.e., considering the fusion of the same two vertices (of H) in G does not alter the Sprague–Grundy value, a contradiction to the fact that G is MNF. □

Lemma 7 (cycle uniqueness) *If G is an MNF, G contains exactly one cycle.*

Proof By contradiction, suppose that G has two cycles. From Lemma 6, both cycles share a common vertex, the root. If the two cycles have also a common edge then they have at least two (distinct) vertices in common, say v, w. But, then, there are, at least, three edge-disjoint paths connecting v, w, contradicting Lemma 5.

Since there are no two cycles with a common edge, G is the sum of two smaller graphs, H_1 and H_2. Since G is MNF, there are two of the vertices in either H_1, without loss of generality, that can be fused so to obtain some rooted graph H_1' such that $g_{\text{SG}}(H_1) = g_{\text{SG}}(H_1')$. But, then

$$g_{\text{SG}}(G) \equiv g_{\text{SG}}(H_1 + H_2) = g_{\text{SG}}(H_1) + g_{\text{SG}}(H_2) = g_{\text{SG}}(H_1') + g_{\text{SG}}(H_2) = g_{\text{SG}}(H_1' + H_2),$$

i.e., considering the fusion of the same two vertices (of H_1) in G does not alter the Sprague–Grundy value, a contradiction to the fact that G is MNF. □

Lemma 8 (bamboo stalks attached to cycle but the root) *If G is MNF, G is a cycle C with disjoint bamboo stalks attached to each vertex of C distinct from the root.*

Proof From Lemma 7, G contains exactly one cycle C. Therefore, any graph attached to every vertex of C is a tree. If one of the attachments were not a bamboo stalk then it would be possible to remove an edge from such a tree, apply the Fusion Principle, and reach a contradiction using the fact that G is MNF (a type of reasoning that was used before).

It remains to be proved that G has no bamboo stalk attached to the ground. If there was one bamboo stalk attached to the ground, G would be the sum of two smaller games: the bamboo stalk and the cycle with other trees. Since G is MNF, the fusion principle could be applied to a pair of vertices in the cycle component without changing its value, and the resulting position would be the same as fusing the same two vertices in the original graph G, contradicting the fact that G is MNF. □

Theorem 3 (validity of the Fusion Principle) *Let G be a (rooted) cycle C with disjoint bamboo stalks attached to each vertex of C, and let G' be the rooted graph obtained after fusing all vertices of C. Then, $g_{SG}(G) = g_{SG}(G')$.*

Proof For visualization purposes, we represent the ground as an extended line and represent G as follows, and use n for the number of edges in C and k for the total number of edges in the bamboo stalks (hence, G has $n + k$ edges),

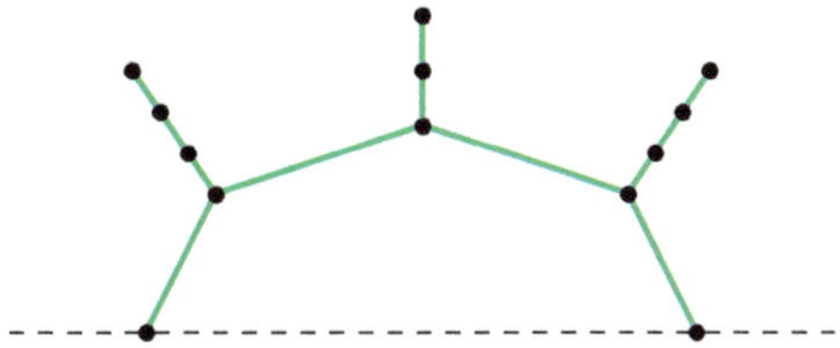

Consider the graph G' that results from G making, recursively, all the fusions until it is no longer possible. This is what is meant by 'fusing all vertices of G'. We will show that $g_{SG}(G) = g_{SG}(G')$ by showing that $G + G'$ is a P-position. The graph G' is the sum of n bamboo stalks and a particular graph, flower shaped, with one vertex and n loops. Graphically, G' looks like the following,

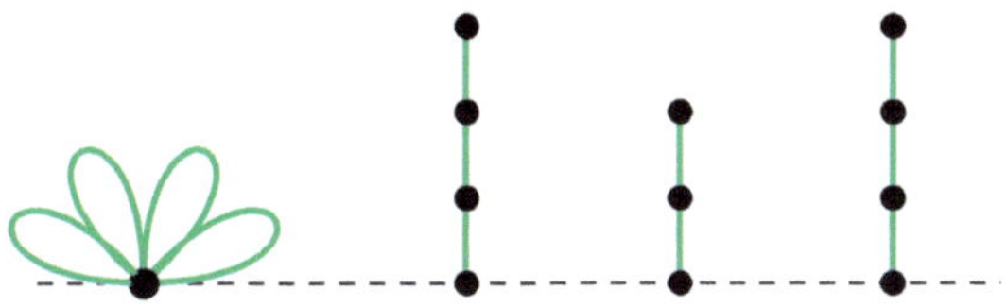

A loop is like a single edge. Therefore, a set of loops on the same vertex is like a set of single-edged bamboo stalks branching on one vertex. Hence, from the Colon Principle, the Sprague–Grundy value of the flower shaped graph is

$$\underbrace{1 \oplus 1 \oplus \cdots \oplus 1}_{\text{n times}} = \begin{cases} 0, & \text{if } n \text{ is even,} \\ 1, & \text{if } n \text{ is odd.} \end{cases} \tag{3}$$

Rename as G' the graph obtained after applying the Colon Principle to the loops. The proof splits depending on the parity of n. First, assume that n is even. The position $G + G'$ looks like the following

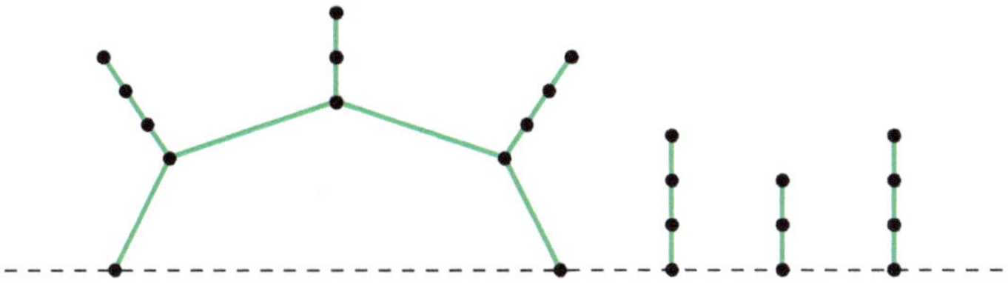

If Player I removes an edge from the cycle then the cycle is broken and turned into two trees. Now, the position of the game is made exclusively of trees and a total of

$$(n-1)+(k)+(k) \equiv (n-1)+2k \tag{4}$$

edges. Since (4) is odd (no matter what k is) then, from the Parity Principle (Corollary 1), the Sprague Grundy value cannot be zero, meaning that the position is winning for Player II (because he is the Next player). Therefore, Player I should avoid removing any edge from the cycle C.

But, if he keeps removing edges from the bamboo stalks, in either G or G', Player II may keep removing the same edge in the opposite graph, until there are no edges left in bamboo stalks and it is Player I's turn to move. Now, Player I will have no alternative but to remove an edge from the cycle, taking the game to a position that is winning for Player II, as we have seen. Thus, if n is even then $G + G'$ is a P-position.

Secondly, assume that n is odd. Thus, $G + G'$ has one extra vertex (and a loop) attached to the ground. The position $G + G'$ looks like:

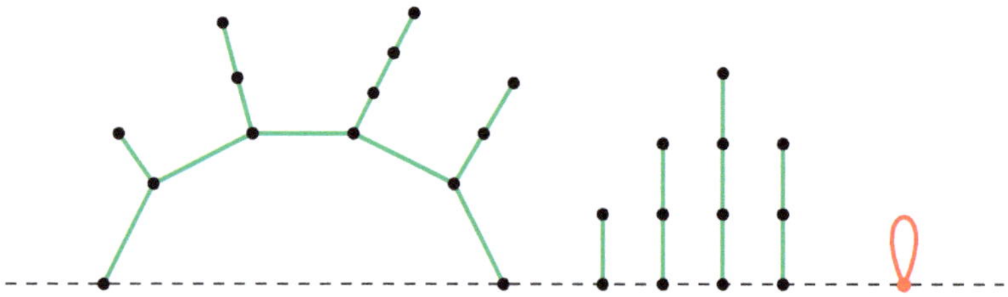

If Player I removes an edge from the cycle then the cycle is broken and turned into two trees. Now, the position of the game is made exclusively of trees and a total of

$$(n-1)+(k)+(k)+(1) \equiv n+2k \tag{5}$$

edges. As before, since (5) is odd (no matter what k is) then, from the Parity Principle (Corollary 1), the Sprague Grundy value cannot be zero, meaning that the position is winning for Player II (the Next to move). Therefore, Player I should avoid removing any edge from the cycle C.

If Player I keeps removing edges from the bamboo stalks, in either G or G', Player II may keep removing the same edge in the opposite graph, until there are no edges left in bamboo stalks and it is Player I's turn to move. Then, Player I's best chance will be to remove the loop because if he removes an edge from the cycle he will surely take the game to a position that is winning for Player II, as we have seen in the previous paragraph. So, let's assume that, at some point, Player I chooses to remove the loop.

We will show that this position is still an N-position, i.e., winning for Player II (the Next to move), by showing the existence of an optimal play that corresponds to removing a carefully selected edge from the cycle C. First, we assign either the label E or O to each of the edges of the cycle, according to the following rule: *adjacent edges are given the same label if and only if there is an odd-length bamboo stalk attached to the common vertex.* Since there are an odd number of edges on the cycle, choose the labels in such a way that there is an even number of E's and an odd number of O's, as the following figure suggests,

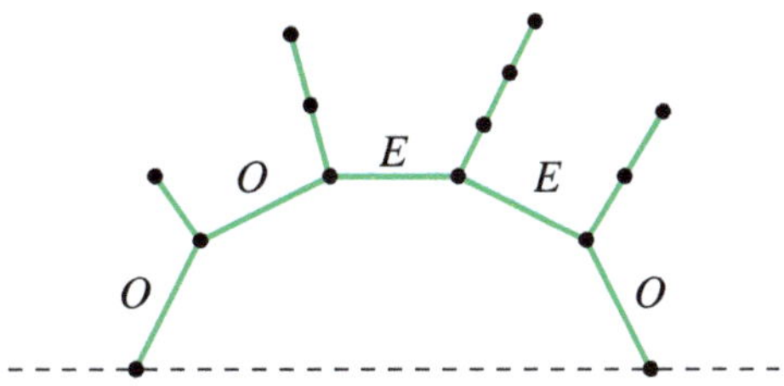

Let e be any O-labelled edge. If e is removed from the cycle then the cycle is broken and turned into two simple (rooted) trees, called them T_1 and T_2. If f is the edge of T_1 furthest away from the root then: **(i)** f received the label E if the vertex that is incident to both e, f has an even length bamboo stalk attached; **(ii)** f received the label O if the vertex that is incident to both e, f has an odd length bamboo stalk attached. Thus, the labeling of the edges along the main path of the simple (rooted) tree T_1 coincides with the labelling of edges in Step 1 of the reduction process described in Sect. 7. The same proof holds for T_2.

Now, consider the simulated position made of the trees rT_1, rT_2, the remaining bamboo stalks cut into half and the edge e put back, attached to the top vertices of rT_1 and rT_2. This position is again a cycle with bamboo stalks attached to its non-root vertices and as many bamboo stalks (with the same sizes) in separate and, since an even number of edges were removed - the E labeled edges and twice as many edges from bamboo stalks—the cycle is still odd. So, the process can be repeated again, and again.

Although it is possible that no edge labeled E is removed, because there is none (which only happens when all bamboo stalks have odd length), the fact that the bamboo stalks are cut into half implies that eventually some bamboo stalk will have even length and, so, some E-labeled edge will appear. Summarising, at some point of this generation process, a position will appear in which the cycle will contain exactly one edge, labeled O, as the following figure shows,

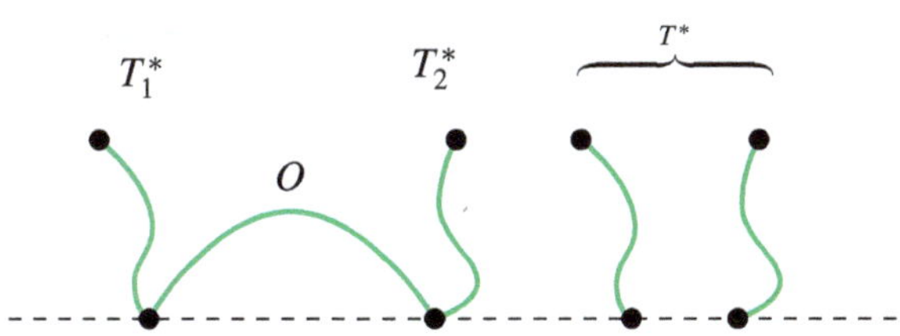

This is an edge of the original cycle that survived throughout this whole process of reductions. The optimal edge is identified! First, observe that removing this single edge takes the last stage of this sequence of reductions to the position with simple

trees T_1^*, T_2^*, and T^* that has Sprague–Grundy zero because, by construction, $T^* \equiv T_1^* + T_2^*$. Now, having identified the edge, we go back to the original position, i.e., $G + G'$ without the loop.

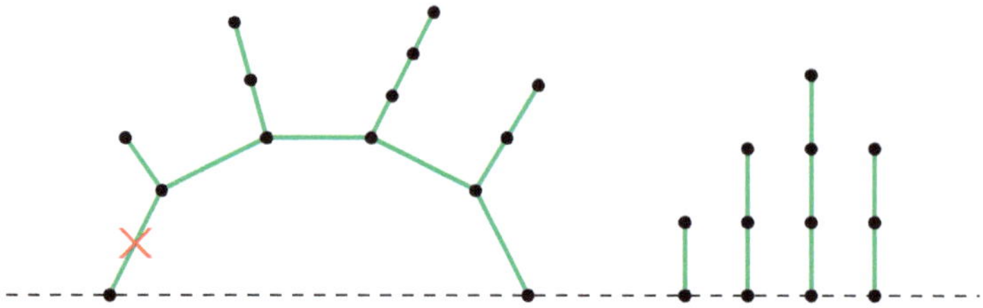

If Player II removes the selected edge from the cycle then the cycle splits into two simple trees. So, the new position of the game is the sum of two simple trees, denoted T_1, T_2, and the individual bamboo stalks, denoted T, which altogether have an even number of edges, because the edges in bamboo stalks appear twice. Now, by construction, the sequence of positions obtained from applying recursively the reduction procedure described is Sect. 7 coincides with the sequence of positions of the simulated reduction process described before in which the selected edge is always the same, except for the fact that the selected edge is not there. Therefore, all the reduced positions will have an even number of edges so that, after a finite number m of reductions, we end up with a position that, as we have seen before, has Sprague–Grundy value equal to zero. Thus, by Corollary 3,

$$g_{\text{SG}}(T_1 + T_2 + T) = 2^m g_{\text{SG}}(r^m T_1 + r^m T_2 + r^m T) = 2^m g_{\text{SG}}(T_1^* + T_2^* + T^*) = 0$$

So, in conclusion, Player II has a winning move which is to remove the edge that was never shrunk during the simulated reduction process. This shows that player I has no winning move in the original game $G + G'$, implying that $G + G'$ is P-position. □

Now, we will conclude that the Fusion Principle holds. If G is MNF then fusing a pair of vertices v, w in the unique cycle of G will take the game to a position G_{vw} such that $g_{\text{SG}}(G) \neq g_{\text{SG}}(G_{vw})$. Let G'_{vw} denote the graph after recursively performing all fusions possible on G_{vw}. Since G is MNF, $g_{\text{SG}}(G_{vw}) = g_{\text{SG}}(G'_{vw})$. On the other hand, G'_{vw} coincides with G', the graph after recursively performing all fusions possible on G. Thus,

$$g_{\text{SG}}(G) \neq g_{\text{SG}}(G_{vw}) = g_{\text{SG}}(G'_{vw}) = g_{\text{SG}}(G'),$$

which, from Lemma 8, contradicts Theorem 3. Finally, let's simplify the Green Hackenbush position below by using both the Colon and Fusion Principles. The purple edges represent a set of edges in which the Colon Principle will be used, and the orange edges are the same but for the Fusion Principle.

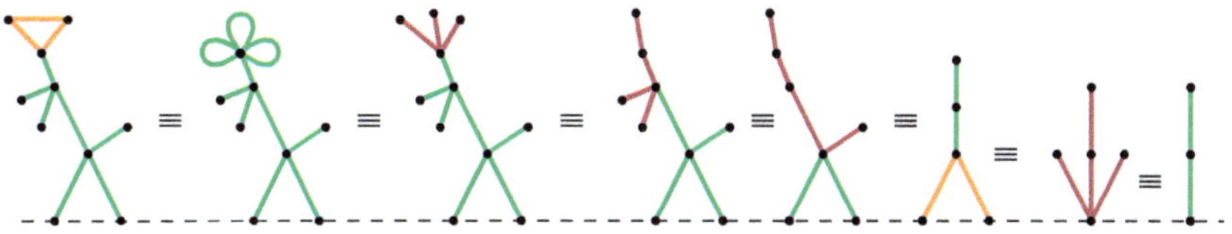

References

1. Hearn, R., A. Franenkel, and A. Siegel. 2014. Theory of combinatorial games. In *Handbook of game theory with economic applications*, chap. 15, vol. 4, ed. Petyon Young and Shmuel Zamir, 811–859. North Holland.
2. Albert, Michael H., and Richard J. Nowakowski. 2009. *Games of no chance 3*, vol. 56. Cambridge University Press.
3. Conway, J. H., R. K. Guy, and E. R. Berlekamp. 2001. *Winning ways for your mathematical plays*, vol. 1, 2nd edn. CRC Press.
4. Bouton, Charles L. 1901. Nim, a game with a complete mathematical theory. *Annals of Mathematics* 3: 35–39.
5. Conway, John Horton. 2001. *On numbers and games*, 2nd edn. AK Peters/CRC Press.
6. dos Santos, Carlos Manuel Ferreira Pereira. 2010. *Some Notes on Partizan Games and Nim dimension*. PhD thesis, Departamento de Matemática, Faculdade de Ciências, Universidade de Lisboa, Portugal.
7. Ferguson, Thomas S. 2020. *A course in game theory*. World Scientific.
8. Grundy, Patrick. 1939. Mathematics and games. *Eureka* 2: 6–8.
9. Guy, R. K. 1989. *Fair Game: How to play impartial combinatorial games*. Comap.
10. John von Neumann, Oskar Morgenstern. 1944. *Theory of games and economic behavior*. Princeton University Press.
11. Nowakowski, R. J. 2017. History of combinatorial game theory. In *Proceedings of the board game studies colloquium XI*, 2nd edn. CreateSpace.
12. Nowakowski, Richard. 2002. *More games of no chance*, vol. 42. Cambridge University Press.
13. Siegel, Aaron. 2003. CGsuite, a computer algebra system for research in combinatorial game theory. https://www.cgsuite.org/ (CGSuite 2.1.1). Accessed 20 Feb 2025.
14. Sprague, R. P. 1935. Über mathematische Kampfspiele. *Tohoku Mathematical Journal* 41: 438–444.
15. Sprague, R. P. 1937. Über zwei Abarten von Nim. *Tohoku Mathematical Journal* 43: 351–354.
16. West, Douglas B. 2000. *Introduction to graph theory*, 2nd ed. Upper Saddle River, NJ: Prentice Hall Inc.

Index

E. A. Pimentel and B. Toni (eds.), *Differential and Algorithmic Intelligent Game Theory*, STEAM-H: Science, Technology, Engineering, Agriculture, Mathematics & Health, https://doi.org/10.1007/978-3-031-97733-6

www.ingramcontent.com/pod-product-compliance
Ingram Content Group UK Ltd.
Pitfield, Milton Keynes, MK11 3LW, UK
UKHW020320200926
475648UK00006B/9
* 9 7 8 3 0 3 1 9 7 7 3 2 9 *